THE ENCYCLOPEDIA OF
ANIMAL
BEHAVIOR

THE ENCYCLOPEDIA OF
ANIMAL
BEHAVIOR

Edited by Professor Peter J. B. Slater

Facts On File Publications
New York

Project Editor: Graham Bateman
Editor: Neil Curtis
Art Editors: Jerry Burman, Chris Munday
Art Assistants: Wayne Ford, Carol Wells
Picture Research: Alison Renney
Production: Clive Sparling
Design: Chris Munday
Index: Barbara James

AN EQUINOX BOOK

First published in the United States of America by
Facts On File, Inc.
460 Park Avenue South, New York,
New York 10016.

Planned and produced by:
Equinox (Oxford) Ltd
Littlegate House
St Ebbe's Street
Oxford OX1 1SQ

Library of Congress Cataloging-in-Publication Data

The Encyclopedia of animal behavior.

   British ed. has title: The Collins encyclopedia of animal
behaviour.
   Bibliography: p.
   Includes index.
   1. Animal behavior. I. Slater, P. J. B (Peter James
Bramwell), 1942–
QL751.E614   1987   591.51   87–8955
ISBN 0–8160–1816–2

Origination by Fotographics, Hong Kong; Alpha
Reprographics Ltd, Harefield, Middx, England.

Filmset by BAS Printers Limited,
Over Wallop, Stockbridge, Hampshire, England.

Printed in Spain by Heraclio Fournier S.A. Vitoria.

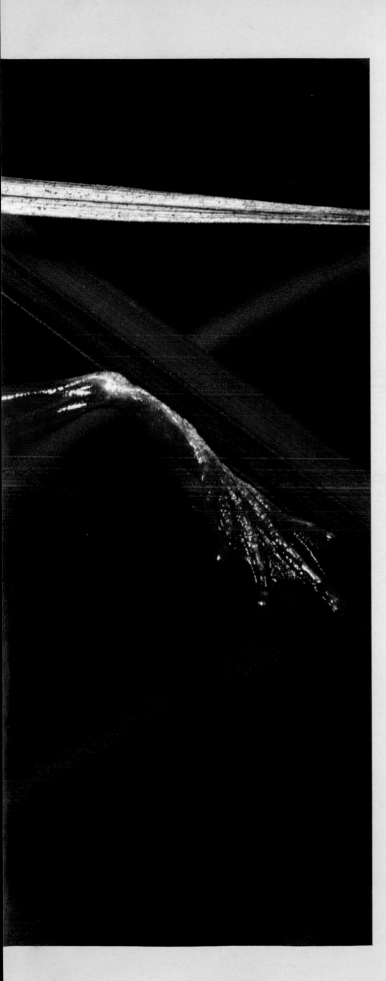

Advisory Editors

Professor Marc Bekoff
University of Colorado
USA

Professor K. Immelmann
Universität Bielefeld
West Germany

Artwork Panels

Priscilla Barrett

Denys Ovenden

Contributors

PPGB	P. P. G. Bateson University of Cambridge Madingley England	PG	Paul Greenwood University of Durham Durham England
MB	Marc Bekoff University of Colorado Colorado USA	TRH	Tim R. Halliday The Open University Milton Keynes England
JTB	John T. Bonner Princeton University New Jersey USA	MHH	M. H. Hansell University of Glasgow Glasgow Scotland
DMB	D. M. Broom University of Reading Reading England	BH	Bernd Heinrich The University of Vermont Vermont USA
JKB	J. K. Burras Botanic Garden Oxford England	IN	Ian Newton Monks Wood Experimental Station England
NRC	Neil R. Chalmers The Open University Milton Keynes England	LPa	Linda Partridge University of Edinburgh Edinburgh Scotland
MD	Martin Daly McMaster University Ontario Canada	TJR	Tim J. Roper University of Sussex Brighton England
RIMD	Robin I. M. Dunbar Cambridge England	DFS	David F. Sherry University of Toronto Toronto Canada
IJHD	Ian J. H. Duncan Poultry Research Centre Midlothian Scotland	PJBS	Peter J. B. Slater University of St Andrews Fife Scotland
ME/JE	Malcolm and Janet Edmunds Lancashire Polytechnic Preston England	PKS	Peter K. Smith University of Sheffield Sheffield England
JBF	John B. Free Rothamsted Experimental Station Harpenden England	CWt	Charles Walcott Cornell University New York USA
JLG	James L. Gould Princeton University New Jersey USA	WW	W. Wiltschko J. W. Goethe-Universität Frankfurt West Germany

Left: two male Red-legged tree frogs (Hyla bipunctata) *fighting (M. Fogden); half-title: Savanna sparrow* (Passerculus sandwichensis) *singing (Dwight R. Kuhn); title page: macaques grooming (Agence Nature).*

PREFACE

To some extent, all of us are experts on behavior. By the way a person walks, or the merest twitch of a muscle in the face, we can tell the mood he or she is in. If we own pets we learn how to interpret their actions: we can distinguish a friendly dog from an angry one, a hungry one from one that wants to be taken for a walk. Yet most people's knowledge ends at about this point because learning about the behavior of animals in greater detail is an extremely time-consuming process. When we enjoy the fascinating wildlife films on television, it is worth remembering that it may have taken hundreds of hours of watching before the points of interest were captured on film. Many animals spend a great deal of their time asleep, and much of the rest quietly foraging, often in dense undergrowth where it is hard to see them. It takes an immense amount of patience to document their activities, as well as care and skill in getting close to them without causing disturbance. Small wonder, then, that most people interested in animals are content to identify them, count them and note down where they are before passing on in search of others. Understanding animal behavior requires a much greater involvement in time and effort.

This book provides a window onto discoveries, made by those who are prepared to stop and watch, about the behavior of animals. At one level, it is about natural history, for the behavior of animals is part of their natural history. But it also concerns science, for understanding animal behavior in all its different facets is a branch of general biology like any other. It requires careful observations and rigorous experiments, in the field as well as the laboratory, if we are able to gain insight into the hows and whys of animal behavior. Those of us lucky enough to be involved in this enterprise know just what a fascinating and exciting area of study it is. New facts about the ways in which animals behave come to light all the time, and, based upon these facts, are intriguing ideas and theories. We hope that these pages will allow us to share with others, professional biologist and amateur natural historian alike, some of that fascination and excitement.

The volume encompasses the whole area of animal behavior. It begins with a broad introduction, setting the subject in its historical context and identifying the sorts of questions with which it is concerned. The subsequent sections build up from the behavior of individual animals to a consideration of the social structures in which they live. The first main section deals with animals as individuals, discussing such topics as the way they find food and avoid being eaten themselves, the shelters they build and the mechanisms of orientation and navigation that enable them to find their way around. The second section concerns relationships between animals and especially how they communicate with one another during the courtship and aggressive encounters which are, for many animals, the stuff of which relationships are made. In the third section, we consider how behavior has come to be the way it is. We discuss its evolutionary origins and its origins within the lifetime of an individual animal, including the way in which inheritance, and factors such as learning, combine to lead to the behavior that we see. The last section concentrates on social behavior. We look at the structures of animal societies and how the social relationships within them develop, as well as at culture and at those interesting cases where different species associate with one another. Finally, the book ends with a glossary for quick explanation of terms that may be new, and a book list to help those who wish to take the subject further.

Every section of the book is split into a series of individual entries on key areas of animal behavior, each written by an authority in that area. The main text provides an overall view of the field as a whole, but the entries are also accompanied by boxed entries in which selected aspects are dealt with more fully. Double page, special feature entries are also included so that particularly interesting topics can be afforded more extended and detailed treatment.

Animal behaviorists seem to delight in regaling others with tales of their discoveries and, fortunately, it is not hard for them to tell their stories without resorting to the jargon that bedevils so much of scientific research. This has been a great advantage in compiling this book and we hope that, from each page, the pleasures of discovery are obvious to everyone. To acquire the individual articles, we have been able to approach a selection of the leading experts in the field, and it has been very rewarding to find how willing they have been to assist. Thus, the articles are definitive accounts contributed by researchers who are prominent in the particular area of study about which they are writing with all the authority that this gives them.

Complementing the text, superb original artwork by several leading wildlife artists together with many color photographs

Spotted hyenas hunting wildebeest (Priscilla Barrett—see pp28–29).

Threat display of an Emperor goose (*Anser canagicus*) defending her nest (Brian Hawkes).

demonstrating the skills of a host of brilliant photographers, illustrate the actions of the animals as never before. To pack so much expertise between the two covers of a book depends upon extraordinary organizational ability and a vision of the end product long before it actually begins to take shape. The credit for this falls clearly upon the shoulders of the editorial team at Equinox (Oxford) Limited and, in particular, on those of Dr Graham Bateman, the Senior Natural History Editor, whose responsibility this book has been. Members of the team have shown meticulous attention to detail in all aspects of the book's preparation, ensuring not only that the facts are accurate but also that they are presented in a most attractive and interesting way. All that remains is for the reader to enjoy and benefit from the results!

Peter J. B. Slater
DEPARTMENT OF ZOOLOGY AND MARINE BIOLOGY
UNIVERSITY OF ST ANDREWS

CONTENTS

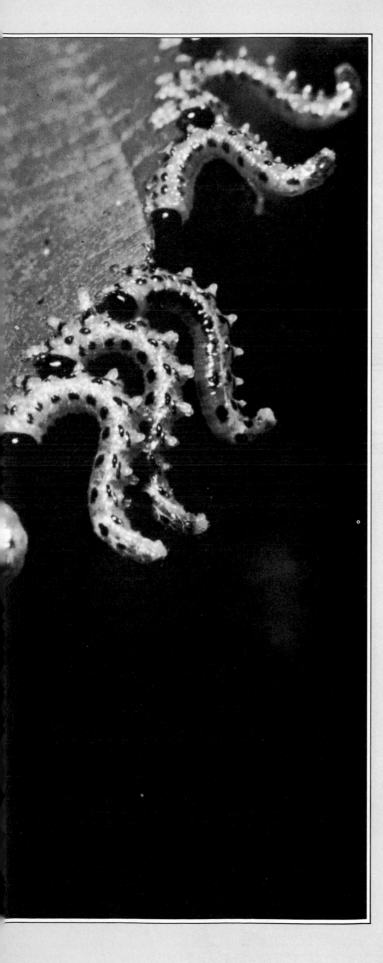

Larvae of the sawfly Croesus septentrionalis *gathered around an alder leaf (Premaphotos Wildlife).*

ETHOLOGY -
The Study of Animal Behavior

*Behavior and early naturalists. . . The Darwinian
revolution. . . Contrasts between the biological and
psychological approaches. . . Tinbergen and Pavlov. . .
Instinct versus learning. . . Behavior in relation to
evolution. . . Key figures in ethology. . . The relevance of
animal behavior to domesticated animals. . . Human behavior
as a branch of animal behavior. . .*

Humans have always been interested in the behavior of the animals that surround them, partly because our species is inherently very inquisitive and partly for good practical reasons. Animals affect the lives of people in numerous ways and knowing about what they do enables us to make the best use of them and avoid their less pleasant effects. As hunters on the African savanna a few million years ago, humans had to understand their prey animals to outwit them, staying downwind and watching their every move as they stalked them silently through the grass ready to fling their spears at the slightest suggestion that their presence was detected.

Vulnerable, naked and defenseless, humans also had to understand that other array of animals, the predators for which they were potential meals, where they hid and how they, in turn, hunted and could be foiled. The knowledge gained in these early times was passed on as a body of folklore and must have been invaluable when people took to a more settled existence and domesticated the animals that were useful as food, such as sheep and cattle, and as helpers in various tasks, such as dogs and horses. Understanding these, our domestic companions, is still an important and useful task for those who study animal behavior and helps us get the best from them and look to their welfare. With the growth of medicine in the past 100 years, another benefit of studying the behavior of animals is that it may help us to understand such crucial matters as the effects that drugs have upon us and the reasons for mental illness. Finally, the subject is important from a conservation point of view. Many habitats are diminishing at an alarming rate through human exploitation and, with this, an increasing number of species are under threat. Only by a detailed understanding of their ecology and behavior is it possible to ensure that populations can be preserved and managed.

Thus, an understanding of animal behavior becomes increasingly important to us. Yet much of the study that has been devoted to the subject would hardly be called practically useful. It has stemmed more from our insatiable desire to understand the workings of the world around us rather than because it was necessary for our health or survival to know the answers. Nature is, quite simply, fascinating, and some of the things that animals do are curious, striking and perplexing, forcing the inquisitive to wonder and to question. Where do swallows go in the winter? Why do they go? How do they get there? Why do bees visit some flowers more than others? Do they learn which is most profitable? And what are they all up to in the amazing melée of their hive? It is towards answering these sorts of questions that the interest of naturalists has turned for centuries and that much animal behavior research is currently directed. Slowly but surely, these efforts have shown that the behavior of animals is far from a random business of curious and inexplicable antics, but can be understood and explained just like other events in the natural world.

Until recently, it was not very clear just how animal behavior should be studied, and there was a shortage of theories against which to test the observations that were made. For example, the Greek philosopher Aristotle collected a great deal of information on animal behavior, but his world was a limited one and many of his answers to the questions that struck him could be no more than speculations, often gloriously wrong. He thought, for example, that bees made honey by distilling dew, their activities clearly being governed by mysterious and astrological forces, for he wrote that they made it chiefly at the risings of the constellations and when there was a rainbow in the sky. But, in many things, he was far ahead of his time.

When interest did return it was among naturalists, such as Gilbert White of Selborne, England who, like several others, was a country vicar with a charge that was not exacting and the leisure to unravel some of God's secrets and the urge to reveal his greater glory. White lived in the late eighteenth century and made meticulous observations of many aspects of natural history although, again, he was limited by lack of evidence on many things. For instance, like Aristotle, he thought it likely that swallows spent the winter in the mud at the bottom of ponds: a not unreasonable idea given the way that they gather in reedbeds before they migrate. He was also more fond of observation and description than of scientific study. But the idea that things were as they were simply because God had

▲ **Understanding nature** was vitally important for primitive man in his battle for survival. Shown here are rock carvings of the animals hunted by bushmen in southern Africa.

▶ **An early theory of Aristotle** was that redstarts (*Phoenicurus phoenicurus*) ABOVE turned into robins (*Erithacus rubecula*) BELOW in the winter because the two species look quite similar and were only to be seen where he lived at different times of the year.

created them that way did not satisfy others; it could explain everything, but helped to understand nothing . . . and there were difficulties, such as fossils. The time was exactly right for the theory of evolution and two men, Charles Darwin and Alfred Russell Wallace, hit upon it almost simultaneously. Biology was revolutionized, and the study of animal behavior with it.

The influence of the Darwinian revolution on how people viewed animal behavior was profound but not, curiously, because of what Darwin himself said about behavior, although he did write quite a lot about the subject. Instead, his impact was a more general one and was primarily based on two points. First, was the fact that he provided a theory relevant to virtually all aspects of living things, including that of behavior. A theory was just what was needed to move the study of animal behavior forward from being a branch of natural history, involving simply the description of what animals did, to being a fully fledged science. A second influence that the theory of evolution had was even more profound. It stated quite categorically that humans were a part of nature, albeit on the uppermost branch of the evolutionary tree. Prior to Darwin, it was possible to argue, with the French philosopher René Descartes, for a split

◄ **Evolutionary links through behavior.** Just as aspects of anatomy can be used to indicate the relationships between species, so too behavioral characteristics tend to be shared between close relatives. For example, most wading birds scratch their heads by bringing their foot up underneath the wing, but plovers and oystercatchers lower the wing and bring the foot forward above it. This is evidence that oystercatchers (*Haematopus ostralegus*) (1) are more closely related to plovers than to other waders, such as the Black-tailed godwit (*Limosa limosa*) (2). However, such similarities may arise by convergence. When they drink, most birds take a mouthful of water and then raise their heads to let it drop down their throats, as with the goldfinch (*Carduelis carduelis*) (3). Pigeons, such as the Wood pigeon (*Columba palumbus*) (4), are exceptions as they are capable of sucking and swallowing without lifting their heads. Some small Australian finches can also drink like this and this must be by convergence as it is known that they are not at all closely related to pigeons.

◄ **The fossil enigma.** Early naturalists who believed that God created all life, found it difficult to explain the existence of fossils such as this ammonite.

► **Hotbed of evolution**—scene from Hood Island in the Galápagos.

between mind and matter, the former being an attribute of intelligent humans alone, while animals were guided solely by their instincts. After Darwin, such hard-and-fast distinctions became difficult to sustain. Humans were animals like any other, even if rather special ones, and human and animal behavior had to be viewed as having arisen through the same evolutionary process. That studying animals might shed light on humans became less outrageous.

In the time since Darwin wrote *On the Origin of Species by Means of Natural Selection and the Preservation of Favoured Races in the Struggle for Life* in 1859, the study of animal behavior has expanded so that it is now a thriving and important branch of biology, usually referred to as "ethology." Behavior is studied by many people throughout the world and its fundamental principles are taught in almost all courses of biology and psychology. Animal behavior today is studied by both psychologists and biologists but their emphases tend to be rather different. Those who call themselves ethologists were usually trained in biology and so approach behavior from the viewpoint of its place in nature. They often study animals in the wild, attempting to disturb them as little as possible and see just how the behavior they show is adapted to their way of life. By contrast, most psychologists are mainly interested in the behavior of people and, when they do study animals, they concentrate on topics such as examining the learning feats of which animals are capable. They tend to study rats or pigeons in laboratory conditions which are very tightly controlled to ensure that the results are consistent. They are not so interested in differences between species as in the rules of behavior common to all of them and so likely to be relevant to our own behavior. But the description above caricatures the two approaches: the ethologist constrained in his hide looking out at the freely moving animal, while the psychologist moves freely about his labora-

tory collecting results from an animal constrained in a small box. Such strong differences were real enough some years ago but today they are not as stark. Many psychologists study learning in a wide variety of species in more natural conditions and, for their part, many ethologists have discovered the merits of laboratory experiments for answering the questions in which they are interested. The distinction is blurring around the edges.

One of the leading ethologists of this century, Niko Tinbergen, summarized the subject by saying that ethologists are interested in asking four different types of question about the behavior of animals. These concern the development of behavior, its immediate causes, how it evolved and the functions it now serves. Two of these questions overlap considerably with those in which psychologists are interested: these are questions concerning the causes of behavior and its development. In asking what causes behavior, we can examine the sorts of stimuli in the outside world that lead to the appearance of certain actions. Does the red breast of a male robin cause others to be more aggressive? Does his song attract females? We can also try to discover what internal mechanisms lead to the behavior. What senses and muscles are involved and what parts of the brain? Do chemicals, such as hormones or the amount of sugar in the blood, affect it? All these are aspects of what causes behavior to occur when it does and are subjects in which both psychologists and ethologists have been very interested. Indeed, the study of the causes of behavior dominated ethology during its great growth period just before and after World War II. This was the time when Konrad Lorenz and Niko Tinbergen began their studies and developed a comprehensive theory of behavior concentrated largely on its causes.

Perhaps the greatest contrast between ethologists and psychologists earlier in this century was in the study of development. Many psychologists had been impressed by the work of Pavlov, the Russian physiologist, who had discovered the conditioned reflex in dogs. They thought that mechanisms, such as those he had demonstrated, could be responsible for the development of much of behavior without the need to trouble about inheritance. The study of learning mechanisms in animals became of great importance to them, culminating in the school of Skinnerian psychology which is largely concerned with rats pressing levers in small boxes to receive food rewards (see p98). The feats that many animals can be trained to do

are certainly remarkable. The approach of ethologists was in marked contrast to this and, to begin with, studying the development of behavior was of much less interest to them than was the study of its causes. Many of the behavior patterns that they studied, such as courtship displays or grooming movements, were rather similar among different animals of the same species. Ethologists tended to assume that this meant they were inherited and appeared fully formed without complex processes of development taking place. Thus, psychologists stressed learning and ethologists instinct and there was no point of contact between them.

As most psychologists now realize, however, inheritance affects behavior in all its aspects. There have now been many studies which show that its influence is no simple matter with some features inherited and others learned. Lorenz himself studied how young birds of species which leave the nest soon after hatching learn to recognize their mothers, the process known as "imprinting," and showed that this could easily go awry if the first conspicuous object they saw after they hatched was a distinguished professor rather than a mother goose. So a very fixed aspect of behavior, that is, to which species the goslings thought they belonged, could be altered if they were

reared by the wrong one; they had to learn what their own mother looked like.

Today, studies of how behavior develops have increased to take on great importance within ethology. They show just what a broad range of processes is involved as the young animal matures and its behavior changes. In some aspects it may need little experience to behave appropriately but, in others, practice, learning by trial and error and copying from other individuals may all have a part to play. Especially where animals live in social groups, the young ones must change their behavior at each stage from the period when they are dependent on their mother to that when they reach adulthood, and must learn about many others in doing so. When you consider the apparent simplicity of a fertilized egg, having more in common with an amoeba than with a human or a monkey, it is staggering to realize that it changes during development to give rise to a social animal of great sophistication and skill.

The two other questions which Niko Tinbergen posed are ones which are peculiar to ethology and of little interest to psychologists because they have their roots very much in biology. These are the questions to which Darwin's theory gave the greatest impetus although, curiously, the thrust of that impetus

took a long time to develop and has only come to fruition in the last few years. It has been an exciting time. The two questions concern the evolution and function of behavior. Evolution is not an easy subject to study directly because it is very slow, and the best we can usually achieve is to reconstruct what happened in the past. But behavior itself leaves no fossils so exactly how it evolved is usually lost in the mists of time. Only in rare cases can we gain insight into what it must have been like from footprints preserved in the primeval mud or the fossilized nests of animals such as birds or termites. The closest to reconstruction that we can normally approach is to look for similarities between related species: if they share a behavior pattern this is likely to be because it was in the repertoire of their common ancestor before they split off from one another. This behavior is then homologous. Caution is needed, however, for sometimes two species may have come to behave similarly through convergence, because the problems they have to cope with are similar rather than through common ancestry. In this case their

▶ **Knowledge changes**—mad March hares. Until recently it would have been thought that these two animals were males. The very latest research has revealed that males and females box with each other as part of their courtship.

Key Figures in the Study of Behavior

As with much else in biology, the field of animal behavior was revolutionized by the work of **Charles R. Darwin** (1809–1882) (BELOW). Prior to his time, it was possible to view animals as primarily instinctive creatures, quite separate from intelligent humans. Darwin's theory of evolution, developed largely from his extensive observations while voyaging round the world on H.M.S. *Beagle*, provided an explanation for how animals come to be so well adapted to their surroundings and good reason for viewing humans as part of nature. In his most important book, *On the Origin of Species....* Darwin devoted a chapter to behavior, but his main work on the subject, in which he

stressed parallels between animals and humans, was *The Expression of the Emotions in Man and Animals* (1872). His own contributions to the study of behavior are, however, slight compared with the impact of his theories on the studies of later ethologists.

Karl R. von Frisch (1886–1982) was born in Vienna but spent most of his life in Germany. He is best known for his discovery of the dance of honeybees with which foraging worker bees inform others of the location of a fruitful food source. He did, however, also carry out important work on the hearing and color vision of fishes. Many doubted his claims that bees were able to transmit to one another information about the direction and distance of food, and some recent experiments have questioned this. Later work, however, notably by J. L. Gould, has demonstrated clearly that bees have an elaborate system of communication.

Konrad Z. Lorenz (1903–   ) is often described as the founding father of ethology. Although several others carried out studies on the subject

between Darwin's and his time, his contribution was more wide ranging. During the 1930s, he developed a theory of animal behavior which stressed its inherited aspects and relative fixity. He based this on extensive studies of his own, largely on the behavior of birds, and the ideas that he developed provided an immense impetus to research in the area. His popular books, *King Solomon's Ring* (1952) and *Man Meets Dog* (1954), show his deep understanding of animals and infected many people with an enthusiasm for the study of their behavior. More controversially, his book *On Aggression* (1966) received widespread attention for his suggestion that aggression is an urge which can only be removed by being sidetracked into harmless activities, such as sports.

I. P. Pavlov (1849–1936) (ABOVE RIGHT) was a physiologist, born in Ryazan in czarist Russia, whose work achieved worldwide recognition and remains an important influence on the ideas of psychologists, especially in the Soviet Union. He is best known for his studies on conditioned reflexes which have formed the

basis of much subsequent work on learning. In his best-known experiment, he studied the salivation of dogs when they are presented with food and showed that, if a bell is rung every time the food appears, the dogs will ultimately come to salivate to the sound of the bell alone. The response to the food is a reflex; that to the bell he viewed also as a reflex, but one conditional upon the animal learning to associate the bell with food. He viewed the building up of such associations as an important part of learning, although he was cautious and did not make such sweeping claims for the generality of his theories as some later psychologists have done.

4

behavior shows an analogy, but not homology. Nevertheless some studies like this have produced very convincing accounts of how behavior has evolved, but it is in studies of function that much of the recent excitement has lain. Biologists use the word "function" in a rather special sense. When they ask what the function of a behavior pattern or some other feature of an animal is, their concern is to know why natural selection has produced it. In other words, what is its survival value or selective advantage to the animals that possess it?

The reason why looking at behavior from this point of view has suddenly become so attractive and interesting is that the full implications of Darwin's theory for behavior have only recently been realized. Twenty years or so ago, it was common for biologists to talk of animals as behaving for the good of their species. This seemed to explain much; the fact that individuals were often seen to help one another and that their aggression was often limited to displaying rather than to damaging fights are two features that fitted in well with this idea. But, more recent work has made it clear that behaving for the good of the species is not at all what Darwin's theory would predict. The most successful animal in evolutionary terms will be that which behaves solely for its own good and for that of its close relatives so that it passes more of its genes into the next generation. Natural selection acts on individuals, not on species; the behavior patterns it favors are those that lead to individuals producing more descendants. Animals may help one another but this is not selfless generosity: recent studies have shown that such cases have arisen because the animals involved are related or because each of them benefits from cooperating. In both these cases giving assistance is advantageous to the individual in the business of passing on its genes.

Studying behavior from this viewpoint, trying to work out how selection has led it to be the way it is and how each of its facets is adaptive to the individual, has become the great "growth industry" of ethology. It is often referred to as "sociobiology," because its accent is on the social behavior of animals. Remarkably, for such a simple theory, just what the theory of evolution would predict and how animals comply with these predictions is still in the process of being worked out.

These then are the four interests of ethologists: in the development of behavior; in its causes; in its evolution; and in the functions that it now serves. They are complementary and each question may be asked about any given aspect of behavior. Some of the best ethology concerns all four of them.

An important point about behavior and how it has been shaped by natural selection is to appreciate that this has happened in a particular environment: bees are adapted to looking for nectar on warm sunny days, ducks to dabbling in shallow water in search of morsels in the mud, lions to crouching hidden in the dry grass and pouncing with alarming swiftness on passing prey. Each has behavior honed by the place where it lives and which we can only fully understand in relation to that place. To understand the causes and development of behavior, we may need to carry out laboratory experiments, but we should not be too surprised if caged animals show aspects of behavior for which we cannot see a reason, because cages were not where the behavior evolved. Studying animals in the wild has much to recommend it scientifically.   PJBS

B. F. Skinner (1904–   ) is the most influential modern follower of "behaviorism," a branch of psychology founded in the United States early in this century which regards learning as of supreme importance. He has studied, in particular, learning by reward (or "reinforcement" as it is often called) and his work has given rise to the school of "Skinnerian psychology." This has sought general laws of learning through the study of animals, usually rats, trained to do tasks such as to press a lever which is arranged to deliver to them a pellet of food. In this "operant conditioning," the amount of a particular behavior pattern, like lever pressing, known as the operant, is raised or lowered by associating it with reward or punishment. Skinner believes the principles emerging from such work to be generally applicable and that even human behavior can be understood in terms of them.

Niko Tinbergen (1907–   ) (ABOVE RIGHT) was born in Holland but moved to England after World War II. His studies of behavior have been made largely in the wild and on a wide range of animals, from butterflies and digger wasps to sticklebacks and gulls. He applied brilliantly simple and elegant experiments to the study of behavior and, through them, was able to gain important insights into the mechanisms underlying what animals do. His *The Study of Instinct* (1951) summarized the theories developed by him and Lorenz and was a milestone in the field. In 1972, together with Konrad Lorenz and Karl von Frisch, Tinbergen received a Nobel Prize in recognition of his contribution to our understanding of behavior.

Making Use of Animal Behavior
Applied biology

Increases in human requirements from animals have coincided with advances in the study of animal behavior. Only very recently has this knowledge been applied to domestic animals.

A traditional way in which an understanding of animal behavior has been put to use is in identifying illness. An early sign of illness is a deviation from normal behavior. A good stockman will suspect illness whenever he sees an animal which is restless when it should be calm, sleeping instead of being active, apathetic instead of being alert or solitary when it should be social.

Knowledge of the behavior of pest species can help in developing control schemes. For example, female Cabbage root flies can be caught in their search for brassicaceous plants by traps baited with mustard oil to which the flies respond. Poison for rats must overcome the fact that rats are extremely wary of new foods, eat only small samples for the first few days and are quick to associate a new taste with a feeling of being unwell (see p99). Some successful attempts have been made to repel bird pests, such as large flocks of starlings roosting in city centers, by playing recordings of their own alarm calls.

An understanding of the feeding behavior of both agricultural and wild species useful to humans can lead to improved performance and yield. Many agricultural species are herbivores but each shows a different type of selective grazing. Horses excrete in certain areas which they then avoid for grazing; sheep like to eat finer grasses, whereas goats prefer certain herbs. Sward management can benefit from such knowledge. Red grouse eat mainly young heather shoots but prefer taller, older heather nearby for cover. Consequently, an optimum habitat can be provided by judicious burning of patches of heather on the grouse moor each year.

In addition to catering for these natural behavior patterns in feeding, use can be made of an animal's ability to learn to perform a response, such as pushing a button, to get a reward. A new way of giving concentrated food to dairy cattle involves a computer-operated food dispenser. A sensing device identifies the cow as it approaches the dispenser by an electronic key worn round its neck; the computer calculates how much food the cow has had in the previous 24 hours and signals for an appropriate amount to be dispensed. The cow has to learn to

◄▼ An emotive issue. Few people would prefer to see farmed animals, such as pigs, housed behind bars BELOW in indoor pens, rather than free-roaming LEFT in open fields. However, the balance of welfare against productivity has to be struck. Modern farming approaches are now recognizing, after thorough behavioral research, that high productivity does not necessarily imply factory farming.

► Happy pigs—a modern pig pen, taking into account natural behavioral needs. The unit of four pens can accommodate four sows with their offspring. Most of the time all the pigs have access to the whole unit. When a sow nears farrowing, she chooses one of the nesting areas and builds a nest with straw she takes from the racks. After farrowing she can be shut in the nesting area for a few days until the piglets can move about easily. Thereafter, they are free to move about as a family group. The layout of the pens and the "furniture" allow natural behavior patterns to take place. For example the sows can build a family sleeping nest each night from where they can see

approaching danger. They can rise in the morning and walk 8–10m (26–32ft) before dunging, as they would do in the wild. There is an area of peat and bark in which they can root, explore and wallow. There are horizontally suspended wooden bars which they can lever upward with their snouts (manipulation of fallen branches is a common activity in the wild). There is also an activity area in which the young pigs can play in straw and a rough wooden post against which they can rub and scratch.

The pens incorporate features from modern husbandry practice which help to increase efficiency. For example, the sow can be restrained behind a farrowing rail for a few days to prevent her crushing the piglets. Also, all the pigs can feed simultaneously without competition from feeding stalls. A creep rail can be added to allow young piglets access to food without the sow getting it.

Breeding sows remain in these pens throughout their reproductive life and their offspring stay there from birth until they reach slaughter weight. A boar spends a few weeks in the unit.

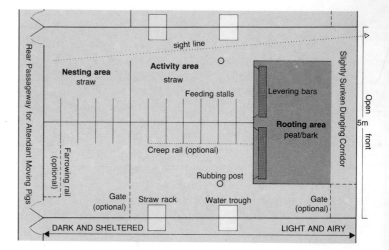

Layout diagram labels: Rear Passageway for Attendant Moving Pigs — sight line — **Nesting area** straw — **Activity area** straw — Feeding stalls — Levering bars — Farrowing rail (optional) — Creep rail (optional) — **Rooting area** peat/bark — Rubbing post — Slightly Sunken Dunging Corridor — Open 5m front — Gate (optional) — Straw rack — Water trough — Gate (optional) — DARK AND SHELTERED — LIGHT AND AIRY

approach the dispenser in a particular way and to space out its visits to obtain a reasonable meal each time. Operant conditioning techniques are also used in training animals such as horses and dogs. In particular, dogs will learn such varied tasks as leading the blind, herding sheep, searching for drugs or explosives and indicating, flushing and retrieving game. In each case, the reward is often only the approval of the handler.

An understanding of reproductive behavior has enabled us to increase reproductive efficiency in farm animals. Visual cues seem to be all-important in sexually stimulating bulls. Thus, bulls kept as semen donors for artificial insemination can be stimulated to mount and ejaculate by intact or castrated males, restrained, non-receptive females or even dummy animals of approximately the right size and shape. The presence of male animals can be used to bring females into heat. Vasectomized rams introduced to ewes prior to the release of stud rams induce the ewes to come into heat at the same time which leads to synchronized, more manageable lambing and a crop of lambs of similar age. Behavior patterns can be used to identify receptiveness in female cattle and pigs allowing them to be inseminated or naturally mated at the correct time. An aerosol spray, containing an artificial pheromone (a chemical substance secreted by an animal which causes a response in another of the same species) similar to the one secreted by the Wild boar, is available which will cause the sow to adopt a rigid, arched-back posture if she is in heat.

In the main, traditional livestock husbandry systems evolved slowly and generally took account of the behavior of the species in question. In the past 30 years, however, the rapid rise of "factory-farming" methods has placed great emphasis on efficiency but seldom was behavior considered. Examples are battery cages for laying hens, crates for veal calves and tether-stalls for pregnant sows. They led to many behavioral problems such as stereotyped movements, excessive aggression, cannibalism and apathy. Now consideration is given to the design of environments for livestock to take account of their behavior. Recent developments are the "getaway cage" for hens and pens which allow pigs to remain in much more natural social groupings and to indulge in species-specific behavior. Similar principles are being applied in zoos to improve animal welfare while allowing the public to see animals engaged in normal behavior typical of the species.

IJHD

Looking at Behavior in Man
Human ethology

Humans are animals, of that we can be sure. And, if we accept the modern theories of evolution, our species emerged from ape-like ancestors only a very few million years ago, adapting behaviorally and physiologically in the process to the changed circumstances of the species and the demands of the environment. But in such a short space of time, we have developed such complex modes of life and culturally influenced patterns of behavior that some psychologists believe it to be difficult if not impossible to regard human behavior in the same light as that of other animals. And, of course, how hard it is to apply the same objective methods of observation to ourselves!

Ethologists who studied the behavior of animals in their natural habitat developed particular methods and approaches and some of these have subsequently been applied to human behavior, creating an area of study known as "human ethology."

Perhaps the most prominent aspect of the ethological approach to human behavior has been the emphasis on observing behavior in relatively natural surroundings. Psychologists usually study people or animals in the laboratory, trying to test very specific theories in controlled conditions. Human ethologists prefer to sit in playgroups, homes, parks, hospitals or other settings where they can watch naturally occurring social interactions and see just what kinds of behavior take place.

Such observations have shown, for example, that there is an attachment system between human parents and babies. After some six months or so, as they become mobile, babies try to stay fairly close to a principal attachment figure (usually their mother or father) by means of following, protesting at separation, and greeting at reunion. For a period, they may also show some wariness or avoidance of adults whom they do not know. This kind of attachment behavior lasts for some two or three years, with the attachment figure(s) serving as a secure base from which the infant can venture out and explore the world—such excursions become more extended in distance and duration as the child gets older.

Much observational work has been done on children in preschool and school-age groups. Some of this has been on facial expressions and non-verbal signals. For example, children show a number of different types of smile, each having a different meaning or corresponding to a different emotional state. Some typical ones are the simple smile, often seen when a child is by himself or herself; the upper smile, usually associated with eye contact and greeting; and the broad smile, seen in laughing and rough-and-tumble play.

Rough-and-tumble play itself has been described in some detail. Apparently more common in boys than in girls, these vigorous forms of play-fighting and chasing are not always approved of by teachers and parents but are enjoyable for the children involved. Aggressive behavior is usually clearly different; associated with frowning rather than laughing, with hitting rather than playfully beating at the other child, and having the intent to hurt. In child and adolescent groups, it has been found that the outcome of aggressive conflicts can often be predicted from a dominance hierarchy; such a hierarchy can be deduced from watching the children, or from asking them to rank others as being the toughest or strongest in their group. Some human ethologists have connected dominance with the idea that the groups have an "attention structure;" children higher in the hierarchy are watched more closely and carefully by those in lower positions.

The naturalistic approach of the human ethologist has thus enriched our knowledge of human behavior, but it has had more to offer than just observational method. Two other important aspects have been an interest in the adaptive significance or functional value of a behavior; and in whether it has been universally observed in different human cultures and even has parallels in some non-human species.

For example, the parent–infant attachment system is adaptive in providing a safe environment for the relatively helpless infant, in which he or she can be sheltered, fed, protected from danger and can learn from friendly adults. Facial signals communicate emotional state and intention; gestures of dominance or submission reduce the risks of actual fighting. The adaptive value of rough-and-tumble play is less certain, but might be to provide physical exercise. Furthermore, all these sorts of behaviors have been observed in a number of different human societies, both preliterate and modern. They may well be

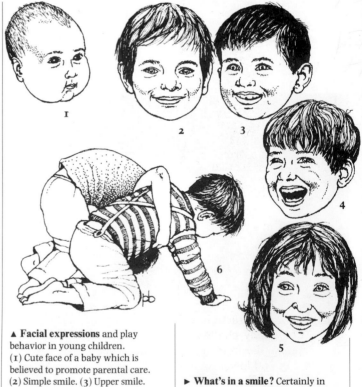

▲ **Facial expressions** and play behavior in young children.
(1) Cute face of a baby which is believed to promote parental care.
(2) Simple smile. (3) Upper smile.
(4) Broad smile. (5) Eyebrow flash.
(6) Children involved in rough-and-tumble play.

► **What's in a smile?** Certainly in this case the broad grin conveys joy to all around.

species-specific characteristics of humans, which are modified but not drastically changed by different cultural conditions. Another example, from adults, is the "eyebrow-flash." This rapid raising and lowering of the eyebrows has been filmed in a great many different societies and seems to be a universal form of greeting.

Some application has also been made of more specific ethological concepts, such as "sign stimulus" and "innate releasing mechanism." For example, it has been argued that certain forms of facial configuration, typical of young babies and of "cute" toy dolls and animals, function to release parental caring behavior (see p88). Similarly, it has been argued that certain stimuli may release aggressive behavior. Psychologists have sometimes argued that there is a number of specific drives which can be released in specified ways. This is very probably too simple a model for humans, however, as it may well be for many lower animals.

Recently, evolutionary biology has brought a more sophisticated theoretical approach to bear on human behavior, again relating to developments in animal studies. This is the application of sociobiological theories to humans, although it, too, has aroused a great deal of controversy. Sociobiologists would not argue that there are fixed drives released in stereotyped ways but rather that behavior is generally chosen so as to maximize "inclusive fitness" (reproductive success of the individual and close relatives). Human sociobiology has attracted a lot of interest among some anthropologists and psychologists, who think that this viewpoint can help to explain such things as human marriage patterns, avoidance of inbreeding, altruism to relatives, and some degree of conflict between parents and offspring. Other social scientists believe that the biologically based universals in human behavior are so weak, that these kinds of behavior can only be explained in terms of particular cultural requirements in different societies. PKS

The Behavior of Individual Animals

Iᴺ this section we deal with animals as single individuals and the problems with which they are faced if they are to survive. Problems that almost all animals have in common are that they must eat and avoid being eaten. But, within this simple statement, there lies a wealth of variety in the sorts of food that different animals eat and in the ways that they cope with the unwelcome attentions of predators. Predators comprise just one feature of the outside world that is hostile. There are many others. From the polar ice to tropical deserts, climate varies enormously and, away from the equator, there are also marked seasonal changes in weather and many other factors.

Furthermore, the earth rotates and so carries its inhabitants through an endless succession of days and nights, which bring with them sharp changes in light, and often temperature. Evolution has generated a remarkable array of behavior patterns to cope with these changes. Animals build shelters of various sorts to avoid the worst vagaries of the climate but, if conditions are too extreme, they may migrate or hibernate. These are dramatic responses to seasonal changes but, less impressively perhaps and over shorter periods, the cycle of sleeping and waking is a way in which animals are adapted to functioning more effectively at one time of day than at another.

◄ **Avoiding being eaten** means for this bush cricket (*Pycnopalpa bicordata*) looking like a leaf, even including brown edges.

FINDING FOOD

Mussels, clams and whales—aquatic filter feeders. . . Grazing animals. . . The problems of hunting for food. . . Choosing the richest food source. . . Defending a territory containing food. . . Learning to eat the correct food. . . Selecting the most profitable food. . . Ants that raise their own food. . . Hoarding food. . . Feeding from flowers. . .

ALL habitats are a rich mixture of the living and the inanimate, of sounds, smells, movement and color. Some of this environment is edible, but most of it is not. Every animal must find the energy and nutrients it requires to fuel its metabolism, grow and reproduce. Finding suitable food among all the things that could potentially be eaten is a very simple matter for some animals but, for others, it is a highly organized process, involving learning, memory and decision making.

Aquatic animals called "filter feeders" select suitable food in a straightforward way. They extract food particles from water carried past them on natural currents, or on currents created by the animals themselves, by filtering the water through their

bodies. Mussels, clams and even some whales feed by filtering the water in which they live.

The food of grazing animals, such as antelope, bison or cattle, spreads out beneath their feet like a carpet and, at certain times of year, is abundant. For these ruminants, finding food is not difficult. It is processing large volumes of their energy-poor diet that requires special adaptations. Because ruminants eat large amounts of food and digest it slowly, feeding and digestion are a nearly continuous process for these animals. The rumen contains a variety of microorganisms that break down cellulose (which is an important part of plant cells) in the leaves, shoots and stems that make up the diet. Although cattle and other grazers may appear to move ponderously over their pastures like mowers, they actually make a variety of choices and decisions while feeding. Some plants are preferred to others, as can be seen by inspecting any pasture, and these preferences usually reflect the concentrations of tannins, alkaloids and other compounds produced by some plants to discourage animals from eating them. In addition, grazing animals adjust their rate of forward movement, how far they swing their head, the rate of biting, and the size of each bite, depending on the quality

▲ **Straining a living** from the sea, the tentacles of this flame scallop (*Lima scabra*) filter food from the water.

◄ **Feeding together**—buffalo and attendant birds. While this African buffalo (*Synceros caffer*) grazes in a swamp, Cattle egrets (*Bubulcus ibis*) hunt for insects it disturbs and oxpeckers (*Buphagus* species) scrutinize its skin for tasty parasites.

► **Nature's farmyard in miniature.** Ants (*Formica obtusopilosa*) milk honeydew from a colony of aphids.

▼ **Tasting with feet.** Blowflies use the fine hairs on their feet to sense chemical features of whatever they are standing on. These chemosensory hairs come in four forms.

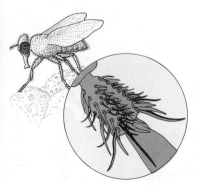

Ants That Raise Their Own Food

Among the most striking in the myriad of adaptations that occur in the insect world are those which occur in ants that grow fungi for food, and others that tend honeydew-producing insects, such as aphids (see BELOW).

It is not clear if the evolutionary origin of fungus growing by ants was simply poor housekeeping by a species that stored seeds and plant material, or whether it arose in ants that lived in decaying wood where fungus occurred naturally. Modern fungus-growing ants, however, devote much time and energy to gathering plant material on which to grow their fungus gardens (the well-known leafcutter ants are fungus gardeners), chewing plant material to promote fungal growth and removing undesirable fungi which compete with those they prefer. Ants use the fungus for food and some species are unable to thrive without it. Most species of fungus-growing ants nurture particular species of fungi which are not found outside ant colonies. Some termites and beetles also grow fungus.

Other species of ants tend livestock, in this case aphids and scale insects that feed on plant sap and excrete a solution rich in carbohydrates, called honeydew. Ants make these insects release their honeydew by stroking them, and then collect the liquid and carry it to the colony. This relationship between ants and aphids is probably mutually beneficial. Some species of aphid have special hairs to hold droplets of honeydew until an ant can collect it; they are also less able to defend themselves than aphids that are not tended by ants. The ants guard their aphids, attacking and driving off other insects that parasitize the aphids. In some cases, the size of the aphid colony, and the growth and maturation of the aphids, are controlled by their ant caretakers. Aphids may also be kept in the ant colony over the winter and carried to new food plants when they begin to grow again in the spring. Some ant colony queens carry a honeydew insect with them in their mandibles during the nuptial flight, when they leave the parent colony, mate and found a new colony.

and amount of food available to them at a particular time.

For other animals, though, finding the right food requires travelling greater distances, and more vigorous searching. It has been estimated that in winter small birds, such as Coal tits or goldcrests, spend over 90 percent of their time searching for food. How do animals like these determine where to look for food, and how do they decide what is food and what is not?

Whether we consider a kingfisher choosing a perch over a stream, or lions waiting at a waterhole, success at finding food depends on where the animal searches. Among the wide range of adaptations that assist all animals in finding food, there are definite preferences about where to look for it. Most woodpeckers look for food along the branches and trunks of trees, and drill holes to extract it, but the flickers of North and South America and the European Green woodpecker are woodpeckers adapted for finding food on the ground. Flickers lack the large reinforced skull and broad chisel-tipped bill of other woodpeckers and have plumages that blend well with the ground. Many woodpeckers have long tongues, but flickers' tongues are especially long, and are used for probing into the burrows of ants, their staple food. Enlarged salivary glands produce a sticky alkaline saliva—sticky to capture ants and alkaline to counter the effects of the ants' formic acid.

Determining the right place to look for food requires the animal to make choices. If a thrush feeding in a rich patch of food, say a berry bush, finds fewer and fewer berries as it continues to search (because it has eaten the ones it found), when should it give up and move on in search of another, as yet untouched, berry bush? Ecologists have attempted to answer this question using theories similar to those used by economists. As these predict, Great tits will remain longer searching through a patch of food when finding a new patch takes a long time. Bumblebees will leave some of a flower's nectar behind if there are many nectar-rich flowers in the area.

Feeding in one spot, like a berry bush, often means that food will not be available there again. Animals ought to be able to remember where they have been eating recently to avoid returning there. The amakihi, one of the Hawaiian honeycreepers, is a bird that extracts nectar from clusters of flowers. Amakihis avoid returning to flowers they have depleted of nectar. Brent geese, which feed in large salt marshes in the Netherlands, crop the vegetation they feed on, but it grows back to the height the geese prefer within four days.

Hoard Now, Eat Later

Since the time of the pharaohs, and probably before, people living in variable environments have known to store food in years of plenty to use in years of famine. Many animals also store food when it is abundant, and later draw on these caches when food is scarce. Rodents and carnivores, birds such as jays and tits and, of course, insects, such as honeybees, collect food which they do not eat immediately but store for use later. We need not imagine that foresight by the animal is involved. Any tendency to store food would be beneficial when food became harder to find and animals with such tendencies would be more likely to survive and breed—natural selection in action.

Small mammals, such as wood mice (1), voles or chipmunks, stockpile grain, seeds and other food in larders in their burrows. Birds, such as jays, nutcrackers (3) and chickadees, store food in thousands of small caches that contain only one or a few items, scattered through their territories. Food is stored in any small cranny that will hold a seed or insect prey. Acorn woodpeckers (2), in western North America, prepare dead trees for use as larders by drilling thousands of holes in the trunk and branches, and then filling these holes with acorns.

Do animals ever use their supply of stored food? Rodents certainly do, especially in early spring when there is little other food to be found because new plant growth has not yet begun and few insects or other invertebrates have emerged from winter dormancy. Food-storing birds also find their caches again and use them. Nutcrackers store food in summer and fall that they use to feed their young, and themselves, the following spring. Birds, such as Marsh tits, store food for a much shorter period of time, collecting their caches a few days after hoarding them. These short-term hoarders probably cache food to keep to themselves rich food supplies that they happen to find. In the wild, a bumper crop of seeds or insects, large enough to provide more than immediate food requirements, is a rare find that attracts many animals.

How do animals that scatter their supply of stored food find it again? Marsh tits, chickadees, nutcrackers and jays have all been found to remember where they hid their food. Not only do they remember the storage sites they used, but some can remember from which sites they have collected food, which sites they have found empty as a result of pilfering by other animals, and what type of food they have hidden at a site. Although these feats of memory seem extraordinary, they are probably just a very striking case of the detailed knowledge of their environment that is possessed by many animals.

▶ **Animals that hoard food.** (1) Wood mouse (*Apodemus sylvaticus*). (2) Acorn woodpecker (*Melanerpes formicivorus*). (3) Nutcracker (*Nucifraga caryocatactes*).

1

Flocks of Brent avoid revisiting areas where they have fed for about four days and then return to harvest the new growth.

When a field mouse finds something to eat, like a sprig of corn, a cricket or some seeds, it probably means there is more of the same kind of food nearby. After finding one food item, many animals search the surrounding area thoroughly, a process called "area-restricted search." Even single-celled animals, such as amoebas, may perform area-restricted search after finding food. If finding food causes them to slow down the rate of movement, or to turn more frequently, they will remain longer in the area where food was first encountered.

Finding food is greatly simplified when an animal can defend, for its own exclusive use, a territory where no other animals of the same species, save perhaps mates or offspring, are permitted to feed. Many nectar-feeding birds defend such a feeding territory, including the African Golden-winged sunbird. Sunbirds defend territories that contain, on average, 1,600 flowers. The territory itself may be large or small in area, depending on how close together these flowers are growing. Biologists have estimated that 1,600 flowers is about the number that can provide the average daily food requirement of a sunbird.

How do animals select what to eat? An animal that fails to select a diet containing all the nutrients and essential vitamins that it needs becomes sick. If, while ill, the animal eats something which corrects the deficiency, it develops a preference for the particular food that made it well again. This has been shown in experiments in which rats fed a diet deficient in the vitamin thiamine were then given food rich in thiamine. After they recovered from the effects of the deficiency, they remembered and preferred the diet rich in thiamine.

A learning mechanism which produces an effect opposite to the one just described allows many animals to avoid eating foods that are toxic or dangerous. The larvae and adults of some insects, such as Monarch butterflies, for example, contain toxic chemicals. Birds that eat these insects associate the appearance of the insect with the illness it produced, even though the illness may not occur immediately after eating. "Learned aversion," as it is called, protects many animals from eating toxic or dangerous foods (and also protects the animals, such as many insects, that produce these defensive chemicals).

Some animals have a head start in selecting safe food. All

▶ **A macrophagous feeder** is defined as an animal that eats large food particles—this Fire-bellied snake (*Leimadophis epinephalus*) seems to be proving the definition despite the efforts of its Harlequin frog (*Atelopus* species) prey to make itself bigger.

young mammals receive milk from their mother. This milk contains not only fats, sugars and proteins but also flavors that are derived from the mother's diet. The milk of cows acquires distinctive flavors from the hay or grain they have eaten (and distinctively unpleasant flavors if they have been into the onions). Experiments have shown that young rodents can recognize the food their mother has been eating from the flavors they detect in her milk. They prefer this food when first exposed to it, with the effect that the first solid food they eat is likely to be a safe and nutritious one.

The way something tastes is often a good indication of whether or not it is good to eat. Humans seem to work this way, but an example from a simpler animal can be used as an illustration. Flies "taste" with chemical receptors on their feet. Blowflies possess only a few types of receptors which allow them to taste salt, water and sugar. If a blowfly puts its foot into something sweet enough to stimulate many sugar receptors, it extends its proboscis and begins to feed. The blowfly uses sweetness to estimate the energy value of the food it has encountered, and has a clear preference for some sugars over others. This example also illustrates that the way something tastes is not only a property of the thing itself, but a property of the receptors an animal possesses. As with all receptors, those that respond to chemicals are adapted to perform specific tasks. Many things in the blowfly's world have no tastes at all because the receptors to detect them are lacking.

In addition to containing different nutrients, food comes in differently sized packets—large and small seeds, for example. Furthermore, these seeds may be easier or harder to manipulate and open before the food they contain can be eaten. If an animal, such as the Coal tit or goldcrest, must obtain the most food it can in a limited period of time, then it pays to select the most profitable food items, the ones that provide the most food for the effort required to handle them. This problem is sometimes called "optimal foraging." Theories of optimal foraging predict that animals will not eat food items in proportion to their abundance in the environment but, instead, will select those items that provide the greatest return of food energy for the time spent handling them. Animals should pass over less profitable food items, if more profitable items are abundant. This is because stopping to handle and eat a less profitable item wastes time that could be spent handling and eating more profitable food. Crows, which break open whelks by dropping them on rocks on the beach, feed in this way. The largest whelks contain the most food, and break most easily when dropped. Small whelks contain little food and must be dropped more times before they break. The most profitable whelks are the large ones, and only these are carried aloft by the crows and dropped. Other whelks are ignored.

One of the most crucial things all animals must do is find food. They are equipped with sensory apparatus, physiological processes, learning and decision-making mechanisms that assist them in this behavior. DFS

Offering a Reward

Feeding from flowers

The living world is a rich and complex web of plants and animals that in various ways are dependent upon one another and upon their environment for survival. To live and grow, animals must obtain enough nourishment and many species achieve this by eating plants or by feeding on the substances, such as nectar, which plants produce to attract them. In turn, the plants benefit as the animals, moving from one flower to another, transfer pollen from plant to plant to ensure fertilization of the flowers and the production of seed.

Flowers have evolved to provide two ways of attracting animals to ensure pollination. Pollen is the functional equivalent of animal sperm because it acts in the fertilization of the flowers. It is a rich source of protein, lipids, vitamins and minerals to pollinating animals, but it is often collected incidentally to nectar, the main means of attracting animals. Nectar is produced specifically to attract pollinators.

Nectar consists of a solution of sugars and it is a highly desirable source of energy, especially to very active small animals that need to replenish their energy supplies regularly. Bees and a few species of social wasps concentrate the nectar to make honey that is stored for times when flowers are not available.

Some animals, such as many insectivorous birds, use just the nectar for food energy and get their protein from other sources, while others use the flowers both as a source of protein for growth and reproduction as well as for energy. Birds which use nectar for energy include hummingbirds (native only in America), sunbirds (Africa and Asia), honeycreepers (Hawaii) and honeyeaters (Australia). Some other birds, such as orioles, white-eyes and parrots, less specialized for foraging from flowers, also occasionally feed on nectar. In America and in the Old World many bats have evolved to feed from flowers. Some of the plants that are specially adapted to be pollinated by bats produce the largest-known volumes of nectar per flower over the flowering period (up to 15ml—some 60,000 times more than is found in many other flowers pollinated by bees).

Bees are totally dependent on flowers for all of their food. The active adults get their food energy from nectar, and the growing larvae obtain their protein from pollen that the adults collect for them. Most other nectar-feeding insects do not get the protein for their larvae from pollen. Butterflies, for example, are made up of the protein that caterpillars have collected by eating large amounts of foliage. Many insects that visit flowers, however, require a protein supplement to mature their eggs. They may eat pollen directly, or they may make proteins from amino acids dissolved in the nectar in small amounts. Some long-lived butterflies, that need to obtain amino acids to make proteins but which have no chewing mouthparts with which to eat pollen, use their proboscis to dip pollen grains into the flower's nectar pools. Amino acids washed out of the pollen into the nectar are then sucked up with the nectar, and are then converted to egg protein by the butterfly.

The nectar of some plants contains poisonous alkaloids. These poisons, which are tolerated by the plant's normal pollinators, may repel other animals, thus ensuring that the reward is left to attract the pollinators themselves.

Most plant species have flowers that are specifically adapted to attract a restricted group of pollinators. In the case of small pollinators this is achieved by having low rewards, so that

◄▲► **The sweet nectar** provided by many flowers is the reward to animals for pollinating them. ABOVE LEFT A male bumblebee (*Bombus pennsylvanicus*) feeds from a thistle in the Arizona desert. LEFT A Greater double-collared sunbird (*Nectarinia afra*) delves into the flower of a King protea in South Africa. ABOVE A Passion vine butterfly (*Heliconius hecale*) probes lantana flowers in the Costa Rican rain forest.

▲ **What an insect sees.** A mallow flower when viewed by ultraviolet light reveals dark lines (nectar guides) which direct a visiting insect to the center of the flower—the site of nectar and pollen. Unlike ourselves, insects are sensitive to ultraviolet light.

pollinators with high energy demands are excluded. For large pollinators the flowers offer high rewards but physically restrict access to these rewards. Access may also be limited by complex flower structure demanding particular foraging "skills" and time of flowering. Bat and moth flowers, for example, open only at night. Specific groups of pollinators are also attracted by signals. Red flowers, for example, are visible against green foliage to birds which have color vision in the red, which most insects lack. Many bee-pollinated flowers reflect in the ultraviolet, which is invisible to mammals and birds.

The pollination resulting from animals (principally insects) feeding from flowers probably accounted for the "explosive" radiation of flowering plants some 100 million years ago in the mid-Cretaceous period from previously wind-pollinated ancestors. Most flowering plants now have highly rewarding flowers, and pollinators seek them out by the bright colors and scents by which they advertise. Using animals to carry pollen, fertilization can occur in plants that are widely dispersed. In turn, this has allowed them to grow in very specialized places.

BH

PREDATORS

O^N a clear day in the northwestern United States, a sparrowhawk suddenly breaks out of its meandering flight, dives toward the ground, captures a small field mouse, and takes off in flight once again. Nearby, a coyote, a member of the dog family and close relative to wolves and jackals, stands still, its piercing eyes fixed on something that has attracted its attention. Suddenly, the coyote begins to walk very slowly and quietly, attempting to conceal its presence from potential food; it then stops, moves its head slightly, sniffs, and then rushes at its prey biting it near the junction of its head and neck. It shakes its head back and forth vigorously, and trots off with a young ground squirrel dangling from its mouth. It then settles down 25m (80ft) away and consumes its meal.

All animals that try to feed themselves primarily on other animals are called "predators" and their meals are called "prey." Predators invest time and energy in locating, capturing

and consuming food items, or they may take advantage of another individual's success and scavenge on already captured prey. And, on occasion, a predator may kill more than it needs at one time and either store unused food in a cache for later use or simply leave it. Red foxes will sometimes cache food in various locations so they do not run the risk of losing all of their food to other animals.

It is important to stress that predators are not simply being bloodthirsty, mean or aggressive. There are "winners" and "losers" in the interactions between predators and prey, but even prey species need fuel for survival that is acquired from eating other living organisms. Typically, prey are vegetarians, however, and do not have to seek out and capture mobile prey. Carnivorous or meat-eating predators also supplement their diet with plant matter and fruits. Red foxes, for example, will often eat apples that drop from trees in the fall.

The ways in which predators hunt for food and then capture and kill it after prey has been located are very diverse. While many predators rely on speed, endurance, power or brute strength to capture prey, others satisfy their hunger by performing sequences of unique and specialized behaviors that are adapted for prey capture. For example, many bats use echolocation or radar to find prey, electric rays eat other fishes that they envelop and then stun with an electrical discharge, shrikes impale prey on sharp branches, and aardvarks are able to trap termites on their sticky tongues. Even Harbor seals may use

◄▲ Predators and their prey.
(1) Aardvark (*Orycteropus afer*) excavating a termite mound; it will then take up a sitting position and extract termites with its long sticky tongue. (2) Red-backed shrike (*Lanius collurio*) impaling a lizard on a thorn; shrikes are famous for their "larders" in which they store food. (3) Cheetah (*Acinonyx jubatus*) chasing a Thomson's gazelle (*Gazella thomsoni*); about half the chases are successful and an average chase is 170m (550 ft) and lasts 20 seconds, rarely exceeding one minute. (4) Grey seal (*Halichoerus grypus*) chasing fish; the diet of this species includes open-sea and bottom fishes (often large specimens) and some invertebrates, such as crabs. (5) A Herring gull (*Larus argentatus*) forcing a Black-headed gull (*Larus ridibundus*) to release food it has taken from a garbage dump; as master scavengers, Herring gull populations are increasing rapidly where they have access to human waste. (6) Noctule bat (*Nyctalus noctula*) homing in on its insect prey, which it would have first detected by the use of ultrasound pulses.

radar when foraging in murky water. Fishers, small carnivorous mammals that live in Canada and along the United States–Canadian border, are adapted to prey successfully on heavily armored, prickly porcupines. Primarily, they attack the face area where there are no quills, and they are large enough to inflict wounds but small and agile enough to jump away, avoiding the slap of the porcupine's tail.

When predators come across mobile prey that are able to take precautions against being captured, the predator must make a series of decisions as the hunt ensues. After prey is detected and identified, the predator must act appropriately, and it usually uses its past experience to increase its chances of successfully capturing and killing the prey.

If prey is sighted and near at hand, a predator may simply rush the prey and hope to outrun it quickly. Cheetahs hunt Thomson's gazelles in this manner, as do other carnivores when chasing small rodents that are unable to seek refuge in a hole in the ground. Cheetahs must get close before the chase because they can only run short distances, usually of less than 350m (400yd). If the prey is detected by a faint odor, barely audible sound, or long-range sighting, the cheetah must somehow get closer to the prey before it begins its actual final pursuit.

For many predators, very slight movements by prey, which are barely sensed by humans, are sufficient to induce it to investigate. Prey detection is also aided by the formation of a "search image." The predator is attuned to certain stimuli that previously were associated with prey. Typically, a predator will stalk its potential, but distant meal by slowly and quietly moving until it is within range where a chase and attack are more likely to succeed than if it simply ran towards the prey when it was first detected. And, of course, the predator itself gives off cues that can be used by the prey to its own advantage;

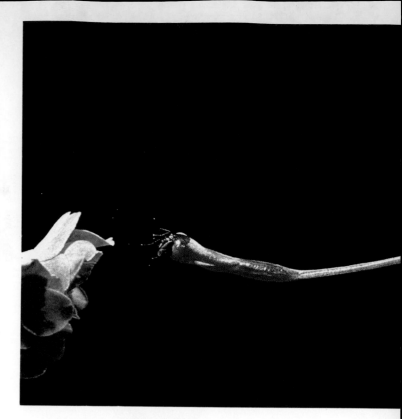

▲ **Sharp-shooter.** A Three-horned chameleon (*Chamaeleo jacksonii*) shoots out its tongue with deadly accuracy to capture a fly.

▶ **Trapped.** A female orb-web spider (*Argiope lobata*) parcels up its grasshopper prey that has blundered into its web.

Detecting Prey

Predatory animals that hunt mobile prey need to be able to locate it with the expenditure of as little energy as possible. Energy spent in this phase of hunting may decrease the ability of the predator to deal with actual capture and manipulation of the prey. Most predators show interesting specializations that are useful either in detecting prey over long distances, in precisely locating where the prey is or in both.

In many cases, non-human predators are able to see, hear, detect fainter odors or sense temperature changes better than humans, and it is necessary to understand the sensory and perceptual world of the animal being studied. Sharks use electrical cues, generated by the nervous system of their prey, to locate

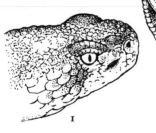

them. Many snakes are able to "taste air" or "see heat." (1) Pitvipers and rattlesnakes can detect a rise in temperature as small as 0.005°C (0.009°F) using heat-sensitive membranes located in pits below their eyes. In this way, they are able to locate a small mouse, for example, when it moves to within about 15cm (6in) of the snake. Many bats (2) are able to hunt at night using echolocation. In this way, they are able to avoid obstacles and locate small flying insects.

Birds and mammals also show sensory specializations that are associated with predatory behavior. Most birds depend on vision and hearing to locate prey but some species, such as Turkey and King vultures, may also be able to use odor cues. Red foxes, which hunt at night, tend to rely most heavily on sound, and Barn owls (3) can home in on the sounds of a mouse rustling among leaves on a pitch-black night with remarkable accuracy.

Brown hyenas (4), which also feed at night and depend on scavenging carrion, appear to depend more on smell.

Of course, all types of cues can work together during a hunt and a successful predator must be able to respond rapidly to changing situations. The cues that are used on a hunt will depend on the time of day that a hunt occurs, the type of habitat in which the predator is located and in which the prey can conceal itself, as well as on prey type.

many prey animals have good all-round vision, for example, and often look up as they feed to watch out for the approach of predators.

The precapture phases involved in a sequence of predatory events often take a lot of time and energy. Instead of stalking, some species hunt by speculation while others merely sit and wait in ambush or build traps in which to capture prey. Some octopuses will regularly close the web between their tentacles over rocks or seaweed when hunting. Then, they feel under the web to see if any prey was taken. If not, the octopus simply continues moving. The predatory larvae of the Water scavenger beetle, a poor swimmer, will sit on vegetation and wait for tadpoles to swim by and then ambush them. Other animals such as hyenas, African wild dogs, lions, vultures and magpies may avoid hunting by feeding on prey killed by other animals, by eating rotten meat, or by finding prey that have died from other causes. It seems that such predators are more attracted to animals that died violent deaths, such as by predation, rather than to prey that died through illness or starvation.

Some predators mimic or resemble another species that is harmless to potential prey. This is called "aggressive mimicry." The predatory Zone-tailed hawk often soars in company with vultures in North America. Small mammals are not afraid of vultures and they do not avoid them. The hawk takes advantage of its concealment among the vultures to dive suddenly from the group and take the prey by surprise.

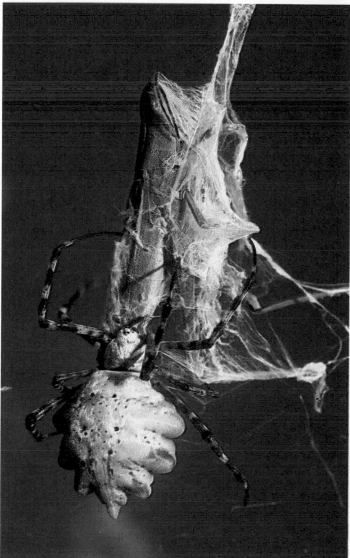

Assuming that a predator does capture its own meal, still more time and effort may be needed to subdue the prey before eating it. Small mammals are frequently pinned to the ground by coyotes, foxes and jackals, and then a killing bite, aimed at the head and neck, is administered. Violent side-to-side shaking of the head often accompanies the bite, during which time the predator's teeth may sink in deeper, the neck of the prey may be broken, nerves and blood vessels destroyed, and the prey disoriented.

In many species, young animals tend to feed on the food to which they were exposed early in life. They have become

imprinted to prefer food items with which they are familiar. In some instances, young individuals watch what their parents or other adults eat and learn to associate certain odors, sounds or sights with specific prey types. They then learn to catch these prey by trial and error. Young and inexperienced individuals are usually less successful than adults in getting a meal especially when particular skills are involved. For example, yearling and two-year old Glaucous-winged gulls are less successful than adults in catching fish.

Predatory behavior takes a lot of time and energy, and we might expect that a wise predator would make every effort to reduce the costs of hunting by selecting prey that are less likely to escape. Indeed, in many instances, young, old, and weak and sick prey are singled out and pursued rather than the healthy individuals that would pose more of a threat to the predator, and decrease the chances for its success.

Predators have difficulty in concentrating on a single individual in a group of prey. Thus, when a predator is sighted, it pays animals that live in groups to shoal, flock, or herd more tightly. Odd individuals within a group may also be subjected to heavier predation than those that are less conspicuous. Predatory birds hunting around industrial centers of England are more likely to prey upon dark moths that alight on light trees and light moths that land on dark backgrounds, than moths that blend in with their environment. Thus, natural selection works to keep prey species as well camouflaged as possible.

Success rates of hunts vary considerably but a few representative ones can be given. In one study of coyotes, ground squirrels were captured 41 percent of the times they were hunted, whereas voles were captured on 19 percent of the hunts during which they were sought.

Among predatory mammals living in the Serengeti Park, where ungulates such as Thomson's gazelles, impalas, zebras and wildebeest are the major prey species, estimated success rates are as follows: cheetahs 37–70 percent; leopards 5 percent; lions 15–30 percent; Spotted hyenas 35 percent; and African wild dogs 50–70 percent. In one study of a pack of 15 or 16 wolves hunting moose on Isle Royale National Park (Michigan, USA), on only six of 120 attempts (5 percent) were moose killed. Other data indicate that Largemouth bass are successful on about 94 percent of their hunts, American kestrels successfully kill rodents or insects on approximately 33 percent of their attempts, and osprey preying on fish succeed on about 80–96 percent of their efforts. It is important to stress that it is to the predator's advantage not to kill all the prey in a population. Indeed, they never do. This is partly, of course, that an arms race is involved: any particularly lethal skill developed by a predator is matched during evolution by an improved anti-predator device on the part of the prey. We would never expect prey capture to be easy. Interestingly, predators do sometimes kill more than they need at one time and either store it for later use in caches or simply leave it uneaten. Such "surplus killing"

▶ **Silent hunter of the night,** a Barn owl (*Tyto alba*) returns to its roost with rodent prey.

may occur sometimes when predators come across prey which cannot escape, as with foxes in a chicken run or confronted with a gull roost on a moonless night when because of the darkness flight is impossible.

When predators feed on prey that is larger and/or faster than themselves, they often hunt in pairs or in larger groups, and success rates tend to increase so that all individuals benefit from the joint effort. When two or more predators, usually belonging to the same species, attempt to find and hunt prey, the process is known as "communal or group hunting." When each individual pays attention to the behavior of other group members, "cooperative hunting" occurs. For example, sharks, dolphins, wolves, and other social carnivores may attempt to herd potential prey. The group spreads out over a wide area and the individuals in the middle of the hunting group approach the prey more slowly than individuals at either end. The prey is eventually surrounded and escape is difficult. Predatory groups of wolves, lions or African wild dogs may also drive prey towards other group members which then ambush it. Group hunting is not always cooperative. Two or more individuals may each be trying to feed themselves without considering the behavior of other group members. Some herons, for example, are attracted to feed close to others because a feeding bird indicates where there is a shoal of fish. But once they are in a group, the birds do not assist each other at all.

While some species such as lions, African wild dogs and wolves tend to hunt mainly in groups, others hunt singly or with other group members, often switching rapidly between the two modes, depending on the prey that is being confronted. For example, when Spotted hyenas hunt Thomson's gazelles or wildebeest, between one and three individuals are usually involved, but, when hunting zebras, the group averages 11 and may include up to 27 animals.

Success rates and group hunting are clearly associated. When lions hunt Thomson's gazelles, a single lion is successful about 15 percent of the time. When two to four lions are involved, the success rate more than doubles to about 32 percent. A similar trend is observed during lion–wildebeest encounters. Likewise, pairs of Golden jackals hunting Thomson's gazelles are more than four times (67 percent) as successful as single jackals (16 percent).

Spotted hyenas also benefit from group hunting. When a single hyena chases and overtakes a wildebeest calf and then is attacked by the calf's mother, the hyena never kills the calf. But, when two or more hyenas are attacked by a single wildebeest mother, they always are successful in getting the calf. The female wildebeest can only deal with one hyena at a time, leaving her calf vulnerable to the other hyena's attacks. Likewise, one hyena is successful about 18 percent of the time

◄ **Cooperative killers.** While one Spotted hyena (*Crocuta crocuta*) distracts a wildebeest mother, other members of the clan chase the easier prey—a newly-born calf. Following a hunt, hyenas feed voraciously—a group of 38 hyenas can reduce a zebra to scraps in 15 minutes.

▼ **Beware of the bears.** Polar bears (*Ursus maritimus*) scavenging from a garbage dump at Churchill, Alaska. Such scavenging is a growing problem where towns have been built near Polar bear migration routes. Since this species is the largest living carnivore, their presence can pose a serious threat to humans.

Polar bears are normally solitary animals hunting alone for their seal prey. It is only at clumped food sources, such as garbage dumps, that they may congregate.

Feeding on Rubbish

Dumps, in which huge heaps of rubbish are deposited, provide food for many different species including invertebrates, rats, jackals, coyotes, bears, dogs, cats, primates and a wide variety of birds particularly several species of gulls.

In one study, over 15,000 individuals representing nine different bird species frequented a dump in Florida, USA, that was only about 150 by 200m (165 by 220yd) in size. Included were about 6,000 Ring-billed gulls, 3,000 Laughing gulls, 3,000 Cattle egrets, 150 Turkey vultures, 1,000 cowbirds and 1,000 starlings. Because so many birds were present, there was intense competition between the species. Herring gulls were dominant over all the others and Cattle egrets dominated all species other than Herring gulls. Vultures typically were avoided when they were at the dump. At another dump site in New Jersey, Herring gulls stole food from Laughing gulls at the dump and when the Laughing gulls flew away. This behavior is called "piracy" or "klepto-parasitism."

The rubbish available at dumps provides a constant and dependable food resource on a year-round basis. Therefore, the survival of adults and young birds after hatching seems to be increased. It has even been suggested that the increase in Herring gull numbers and the expansion of their geographic range are related to food available at dumps.

Another factor that requires careful attention is the location of dumps. If a dump is located near an airport, which many are, birds' presence may be hazardous to aircraft flying nearby. Because so many birds are attracted to dumps, it would be best to place them at considerable distances from airports and out of the flight paths. Or, it becomes necessary to cover the rubbish immediately and constantly to discourage the birds.

Dump management may also influence competition between Herring gulls and Laughing gulls. When bulldozers mash the rubbish, it is advantageous to Laughing gulls. Bags are broken open, more food is available and there is a decrease in competition between the two species. Furthermore, Laughing gulls are smaller and more agile than Herring gulls, so that they can feed between the moving bulldozers. Therefore, patterns of dump management clearly favor Laughing gulls in their food-related battles with Herring gull pirates, and populations of Laughing gulls may increase.

---

when it hunts wildebeest calves whose mothers do not attack it but, in the same situation, two hyenas have a success rate of about 32 percent.

Group living may also be important for food defense. For example, in one study conducted in northern Wyoming, USA, it was found that when coyote pack members were successful in defending elk carrion from intruding non-pack coyotes, an average of 2.6 pack members were involved in the interactions. When pack members lost and intruders were able to gain access to their food resources, only 1.3 pack members partook in the encounters.

In studying predators, we are able to learn how different animals satisfy their hunger, and also how anatomy, physiology, ecology and behavior are all tied together. It is important to stress that predators are not being mean or aggressive when they hunt and satisfy their need for food and the energy that a meal provides. Non-human animals, like their human counterparts, need to eat and many of them have unique adaptations which mean that they can only survive by eating other animals.

MB

Tables Turned
Animal-eating plants

All animals, directly or indirectly, rely on plants for their nutrition; from the grazing herds of the African plains (with their attendant predatory carnivores), to sap-sucking aphids and the wheeling swallows feeding on their winged migratory forms. The relationship between animals and plants is, however, interdependent and, in like manner, land plants directly or indirectly take nutrients from the soil via the products of animals and microorganisms.

In keeping with all green plants, predatory plants obtain carbon and oxygen from the atmosphere. They are, however, able to short circuit the path by which other plants obtain certain nutrients from the soil. To achieve this, nature has provided them with an astonishing variety of ingenious traps by which they are able to attract, ensnare and use as a food source, a wide range of animal prey. By taking nitrogen, phosphorus and possibly other minerals and vitamins from the very bodies of the animals they catch, predatory plants supplement their nutrition and are able to flourish in habitats where other plants would be unlikely to survive. Contrary to popular belief, such habitats are not confined to acid boglands; they also encompass tropical rain forests, cloud forests, savanna and even semideserts. In all these situations, a shortage of nitrogen may be the common denominator and this could be the catalyst in the evolution of predatory plant systems. Among flowering plants, six families consist entirely of carnivorous species; they are: Byblidaceae, Cephalotaceae, Droseraceae, Lentibulariaceae, Nepenthaceae, and Sarraceniaceae. The wide botanical separation of these families and the geographic isolation of, for instance, the Australian Byblidaceae and Cephalotaceae, indicate that the carnivorous habit has arisen more than once in the course of evolution.

The complex trapping mechanisms can be of two main types, either active or passive. The best-known active trapper is the Venus's-flytrap (*Dionaea muscipula*) from the boglands of North and South Carolina. This vegetable gintrap excited the curiosity of Charles Darwin and he described it as "one of the world's most interesting plants." Indeed, we can only marvel at the forces of evolution which have molded a single leaf into a twin-bladed trap, armed with marginal teeth and resembling so much an open clam awaiting its prey. The blades of the trap are united by a hydraulic vegetable hinge to allow so rapid a closure of the blades that luckless insects which have entered the trap have little chance of escape.

Insects and other invertebrates attracted by color and nectar-like secretions, eventually blunder into trigger hairs positioned towards the center of each blade. The closing of the trap on suitable prey results in the build-up of considerable pressure and this usually leads to the death of the prey by crushing. Digestive enzymes are then secreted from glands on the inner faces of the trap and the products of digestion are finally resorbed for the benefit of the plant.

A tiny, though very similar, active trap system is found in the free-floating Waterwheel plant (*Aldrovanda*) which is related to Venus's-flytrap and ensnares and feeds on small forms of aquatic animal life. This rootless water plant has a wide distribution throughout the subtropics of the Old World.

The active trappers also include the sundews (*Drosera* species) and the butterworts (*Pinguicula* species). Sundews are provided with stalked glands on the upper surfaces of the leaves; these secrete a glistening, gummy fluid or mucilage which is one of the stickiest substances in nature. Insect prey, attracted by color and nectar, are caught up in this minefield of mucilage and, in their struggles to escape, more stalked glands curve towards the areas of excitation ensnaring the prey still further. This movement, though slow, is positive and it may readily be detected by time-lapse photography. The plant then makes use of the soft parts of the insects by digesting them with enzymes from glands which also resorb the resulting "soup."

Butterworts ensnare their prey by sticky secretions accompanied by an inward rolling of the leaf margins, and this again is a slow but clear movement and warrants the inclusion of

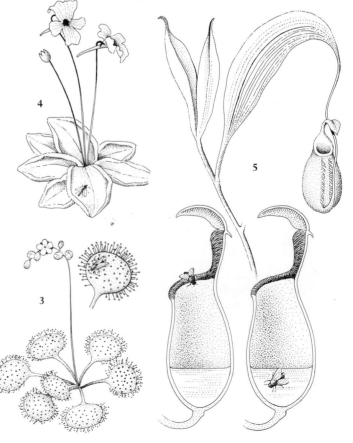

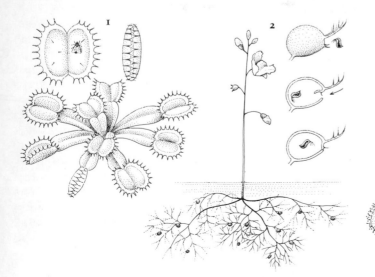

► **A sticky end**—death of a damselfly. A Blue-tailed damselfly ensnared on the sticky tentacles of the Common sundew (*Drosera rotundifolia*).

◄ **Carnivorous plants.** (1) Venus's-flytrap (*Dionaea muscipula*), the toothed leaves of which snap shut on the prey. (2) Bladderwort (*Utricularia* species) in which small prey is sucked into submerged bladders. (3) Sundew (*Drosera* species) where insects become trapped on sticky leaf tentacles. (4) The butterwort (*Pinguicula* species) is another passive catcher of prey on sticky surfaces. (5) Pitcher plants (*Nepenthes* species) produce deep pitchers containing fluid in which insects drown and are digested.

these plants within the category of plants that use active mechanisms.

The most rapid plant movement known to science occurs in the carnivorous aquatic bladderworts, a group of widely distributed species in the genus *Utricularia*. Here, the traps are small bladders provided with elastic walls which become suddenly distended by the inrush of water as the hinged door opens. Free-swimming animals, such as *Daphnia* and rotifers which release the door by colliding with strategically positioned trigger filaments, are swept in. The trap is reset when the vacuum is restored by the removal of water from the bladder.

Predatory plants with passive traps include the pitcher plants. In these the bottle-like traps may be an extension of the leaf midrib, as in the genus *Nepenthes* of the eastern tropics, or a completely modified leaf typified by the North American *Sarracenia*, the Australian *Cephalotus* and the South American sunpitcher *Heliamphora*. Insects attracted to the pitchers by colors and sweet secretions eventually fall into a digestive fluid at the base of the pitcher. Here they quickly drown and are digested. Bacteria may also play a part in the breakdown of the prey of many pitcher plants.

Winged and crawling insects are the principal fare of predatory plants but an extensive variety of other animals have been found in the traps of *Nepenthes* and *Sarracenia*, including small frogs and mice, while the larger bladderworts are known to catch tadpoles and fish fry. JKB

PREY DEFENSE

Color matching in partridge and spiders. . . Counter shading in deer and fish. . . Simple camouflage in insects. . . Warning colors in wasps. . . Mimicry—Müllerian and Batesian. . . Escape strategies—fleeing, distraction, mobbing. . . Retaliation. . . Bats and moths. . . Changing color. . .

EXPLODING into a short and rapid sprint, achieving speeds of up to 90km/h (56mph), a hungry cheetah on the open plains of southern Africa tries to satisfy its need for food by capturing, killing and eating a Thomson's gazelle. More often than not, the hunt fails, the gazelle escapes to live another day and the cheetah, panting and exhausted, must go hungry. As a result of striving to capture enough food over many generations, the cheetah has evolved excellent senses, the power to run for short distances more quickly than any other mammal and formidable teeth and claws for killing. But the gazelle is ever watchful, highly maneuverable and is no sluggard either.

Most animals are in potential danger of being eaten by another animal at some time in their lives but they have all evolved ways of diminishing this danger. A widespread method is to avoid being seen. Many predators hunt by eyesight so that a prey which is well camouflaged ("cryptic") has a good chance of being overlooked. The simplest form of camouflage is color matching: the partridge and wolf spider are brown and live where vegetation is sparse and there is much bare soil, while grasshoppers and caterpillars, which eat leaves, are often green. Many predators, however, are able to detect prey, even if it matches its background, by the shadow on the lower part of the body, or by the characteristic body contour. It is easy to see a green ball on a lawn because the lower part is shaded and its outline is circular. The sunlight striking down on a solid, evenly colored object, throws the lower part into shadow. In animals, such as deer and many fishes, this effect is counteracted by the belly being paler than the back, a phenomenon known as "countershading." When such an animal is viewed from the side with top lighting, its solid appearance is transformed into that of a uniformly hued flat object.

It is much harder to distinguish the shape of a blotched animal on a mottled background than when both are evenly colored. If the patches of different color are bold, they may disrupt the outline. The Ringed plover is strikingly patterned with patches of brown, white and black but, on a stony background containing these three colors, it can be very difficult to see; its brown head patch circled with white seems to resemble a stone that is quite separate from its brown back. On the large caterpillar of the Citrus swallowtail butterfly, black marks break up its outline, even when it is resting on the top of a shiny lemon tree leaf.

Simple camouflage, by matching the background, can be effective but there are other animals with a precise resemblance to a particular inedible part of the environment. Many looper caterpillars and stick insects and some praying mantids closely resemble individual twigs or sticks, while certain bush crickets and leaf insects are precise mimics of green leaves, complete with apparent veins. The praying mantid, *Phyllocrania paradoxa*, resembles a dead, shrivelled brown leaf, and several

Matching the Background

An animal that is camouflaged will become conspicuous if its background changes in color, but some species can change to match their surroundings. Hares, stoats and ptarmigan in northern Europe are brown in summer but molt in the fall to become pure white in winter. They are, thus, well camouflaged at all seasons except possibly for a short time in spring and fall. In the North American swallowtail and in the European Small white butterfly, pupae which overwinter on dead brown plants are usually brown whereas those that pupate and emerge in summer are more often green. Coloration is partly determined by day length: caterpillars reared with only eight hours of daylight become brown, whereas those reared

with 16 hours of daylight may become green or brown. In the swallowtail, summer-reared caterpillars become brown on a rough surface but green on a smooth surface, so that pupae on different backgrounds will be camouflaged.

There are other animals which can change color much more rapidly without having to molt. In these, the color pigment is in individual cells which are under the direct control of hormones and nerves. Flounders and some crabs can vary their hue over a few days to match the environment. The chameleon is renowned for its ability to change color in a few seconds. It is normally camouflaged on its background, however, and, when handled, goes dark in a startling display.

◀ **In shape and color,** the Papuan frogmouth (*Podargus papuensis*) resembles a branch even to the peeling bark.

▲ **Mind the frog.** This Horned frog (*Megophrys nasuta*) from the Malaysian rain forest resembles a dead, brown leaf and thus avoids being seen by predators.

▼ **Foul mimic.** Bird droppings are a common sight on vegetation and caterpillars, such as this one of the butterfly *Papilio aegius*, gain protection from predators by mimicking such droppings.

▷ **Uncanny resemblance** OVERLEAF to a grass leaf makes this Towered or Nosed grasshopper (*Acrida hungarica*) virtually invisible to predators.

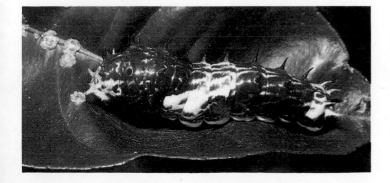

frogs and toads from South America are flattened and mimic dead brown leaves on the forest floor. These stick and leaf mimics may even be ignored by a predator when they are not on a background they resemble. There is even a fish, *Lobotes surinamensis*, which lives in mangrove swamps of South America and rests on its side just below the surface of the water. Here it mimics a mangrove leaf that has fallen in the water, and it has brown blotches resembling blemishes on the leaf. The principle of being visible but not recognized as edible is carried still further by some caterpillars and spiders that rest on the top of leaves and resemble feces dropped by a bird.

The essential principle of camouflage is harmony with the background, or some part of it, but some animals contrast strongly with their surroundings. Sometimes, these bright colors are signals related to courtship and territoriality (see p80). In other animals vivid colors warn potential predators that, if they attack they will have an unpleasant experience. Wasps and hairy caterpillars are familiar examples of animals with warning colors that advertise their distasteful nature or ability to hurt.

A few predators have an innate aversion to specific warning color patterns. Examples among birds are the Great kiskadee and some species of motmot from North America which will avoid sticks banded with red, black and yellow that simulate

poisonous coral snakes even though they have never seen similar objects before. Most birds, however, have to learn by trial and unpleasant experience, that animals, such as yellow-and-black wasps, are nasty. Because inexperienced young birds are reared every year, there must be considerable loss of wasps as each new generation learns the characteristic coloration. So warning coloration only occurs among species that are fairly common.

In a rare species, too high a proportion would be killed by inquisitive predators, so it would pay to be cryptic and so not be found. Warning colors are also most common in gregarious species in which members of the group are related. A bird learns the pattern of a wasp and its association with unpleasant taste by killing one, but that death only benefits the wasps that survive. Wasps live in colonies, however, in which all individuals are related and so they have many genes in common. Therefore, the surviving wasps are likely to carry many of the genes of the one that was killed which, by its death, can enhance the survival of the same genes in its relatives.

In some cases, two or more species of animals look almost identical, having the same brightly colored patterns. If both species are unpleasant, this is "Müllerian mimicry," and both benefit because a predator coming across an individual is more likely to have experienced the pattern before, and so avoid it. In other cases, however, one species may be a perfectly palatable mimic of another warningly colored distasteful species. In this "Batesian mimicry," the mimic gains protection because predators may avoid attacking it when they have learned the pattern of the model. The model gains nothing.

Wasps are common warningly colored animals throughout

▲▶ **Pretending to be dangerous.** The patterns of coloration of the harmless King snake (*Lampropeltis triangulum*) RIGHT of Central American forests closely resembles that of the poisonous Coral snake (*Micrurus nigrocinctus*) ABOVE. Such mimicry of a poisonous animal by a harmless one is known as Batesian mimicry.

◀ **Defenseless cheetah cub.** Some harmless edible animals deter predators by appearing to be really dangerous. This bluff may be an apparent increase in size and ferocity, or an imitation of a dangerous animal. Cheetah (*Acinonyx jubatus*) cubs, instead of being camouflaged, as occurs in leopard cubs, are quite conspicuous with dark bellies and pale backs. They apparently mimic the Honey badger (*Mellivora capensis*), a small but very aggressive carnivore which fearlessly attacks predators.

the world. In Europe they are usually banded with black and yellow, and there are several species of solitary and social wasps which are Müllerian mimics, all capable of stinging if attacked, being nasty to eat, and all sharing the same pattern. There are other insects, such as hoverflies, which are black and yellow but are perfectly edible to birds; these are Batesian mimics which derive protection from their similarity to the wasps. In the Americas there are several species of harmless snakes which are Batesian mimics of the poisonous coral snakes, and the orange Viceroy butterfly is a palatable Batesian mimic of the Monarch butterfly. Not all monarchs, however, are distasteful; those that fed as caterpillars on poisonous species of milkweed are themselves nasty while those that fed on harmless species of plant are perfectly edible, so it is possible to have

Batesian mimics and models belonging to the same species.

Camouflage, warning coloration and mimicry are all excellent defenses, but they do not always work. Occasionally, a predator may find a cryptic animal or be so hungry that it is prepared to eat one that is normally too unpleasant for it to touch. Thus, many animals have additional defenses which protect them when confronted with a hungry predator. Some animals withdraw to a safe place: fanworms into their tubes; snails into their shells; and rabbits into their burrows. Alternatively, an animal may try to escape by speed, either running, flying or swimming, and this will succeed provided that the potential prey can move faster or has greater stamina than its pursuer. The outcome of a chase often depends on how close the predator gets before it is detected by the prey. A cheetah

◄ **Hissing threat.** When threatened chameleons, such as *Chamaeleo namaquensis* shown here, inflate their bodies and expose the brilliant lining to their mouths, while hissing. Throughout Africa these harmless animals are considered highly dangerous by many people.

► **Dangerous and letting it be known.** Many caterpillars are covered in distasteful or poisonous spines which makes eating them unpleasant. With such protection it pays to advertise your presence by bright coloration. Shown here are caterpillars of a silk moth (*Automeris* species).

Battle of the Air Waves

The relationship between bats and moths is one of the most fascinating examples of a predator–prey interaction. Moths are a group of insects which fly at night when predators which hunt by sight cannot see them. At a very early stage in their evolution, bats developed a sonar mechanism which enabled them to fly in the dark without bumping into obstacles. It has gradually changed through evolution so that, today, bats can detect and capture flying insects, especially moths.

The early moths had no ears, but the intense predation by bats would have favored any moth that had some structure that would vibrate when stimulated by the call of a bat so

that it could take evasive action. Moths from several different families evolved simple "ears." Bats can fly much faster than moths, however, so a moth will not escape from a bat simply by detecting it and turning away. Bats can detect moths of reasonable size at about 5m (16ft) distance, but moths can detect bats at perhaps 40m (130ft). Thus, if a moth hears a bat at a considerable distance, its best strategy is to turn and fly away as fast as it can (1) or it may be captured (2). It may never come within the range of detection of the bat. But, if the bat is much nearer, this strategy will be useless because of the bat's superior speed, so moths respond to louder, closer

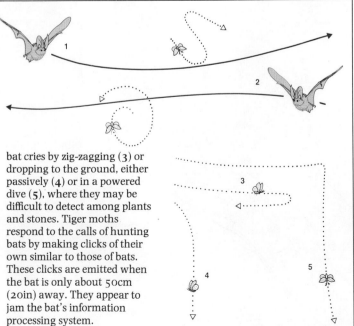

bat cries by zig-zagging (3) or dropping to the ground, either passively (4) or in a powered dive (5), where they may be difficult to detect among plants and stones. Tiger moths respond to the calls of hunting bats by making clicks of their own similar to those of bats. These clicks are emitted when the bat is only about 50cm (20in) away. They appear to jam the bat's information processing system.

can outrun any antelope, but it has no stamina and gives up exhausted if it does not capture its prey within about 400m (440yd), whereas a pack of African wild dogs has less speed but tremendous stamina for a long chase. So antelope tend to adjust their escape behavior to each particular predator: Thomson's gazelle will flee from African wild dogs at 500 to 1,000m (550 to 1,100yd) range whereas they permit the cheetah to approach to 100 to 300m (110 to 330yd) before fleeing. To flee at a greater range from a cheetah would be a waste of energy. Even clams have different responses to different predators. When attacked by most predators, a clam's response is to close the two shells tight and wait until the danger passes. But, when attacked by starfishes many species of clam repeatedly flex their foot and leap away. They recognize the starfish by the chemistry of its tube feet and, because starfishes can open bivalve shells, the clam will have a better chance of

escaping if it moves away instead of closing its shells.

If an animal cannot escape outright from a predator, it may confuse its pursuer and make capture more difficult. Moths, such as the Yellow underwing, have bright colors on the hindwing, called "flash colors." Normally, this insect is cryptic when at rest but, if it is found, it flies quickly away exposing bright yellow flashes on its hindwings as it goes. When it lands the yellow apparently disappears as the wings are closed, and a predator is left baffled. Flash colors occur in many insects and in some tree frogs which have the vivid colors on the inside of the legs, visible when they jump, but hidden when they land. In other species a predator may be deceived into seizing a relatively unimportant part of the prey. Insects, such as the brown butterflies, have small, eye-like spots, and there is evidence that birds direct their pecks at these rather than at the butterfly's head. If the insect is caught by the wings, it may still escape, albeit with a torn wing but, if it is seized by its true head, it will almost certainly be killed. Some tropical blue butterflies have false eyes and also false antennae which divert attacks to the wrong end of the animal. Bluff also occurs in animals such as the American opossum and some stick insects which lie motionless as if dead and can be prodded by a predator without showing any signs of life. Some predators only make a killing attack on prey that moves so that this strategy may protect the prey.

In some instances, the prey's behavior appears to attract the predator's attention. Sometimes, small birds gather round a hawk or owl and "mob" it by conspicuous movements and much noisy calling. It seems strange behavior because we would imagine that a hawk could easily kill any one of the birds. They usually keep out of reach of the predator, however, which, in any case, is not in a good position to attack because most hawks and owls catch prey while in flight and are not agile enough to capture a small bird just a few meters away. Mobbing possibly informs the predator that all potential prey in the vicinity are vigilant so that it would be futile to attack. Mobbing birds may also annoy a predator so much that it flies away from the area. Some birds, such as the Pied flycatcher, mob more frequently when nestlings have fledged, and here the behavior may divert attention away from the defenseless young to the much less vulnerable parents. Even more remarkable methods of diverting the predator's attention from the chicks occur in wading and gamebirds. The parents often flutter away from their young in a way which suggests that they are unable to fly. Such broken wing distraction displays lure predators away from the chicks and, if the predator gets too close, the parent birds simply fly away.

The final defense of most animals, if all else fails, is to retaliate with teeth, horns, claws or whatever weapons are available. Bees sting, sticklebacks erect their spines, millipedes exude repellent secretions from glands along the body, some caterpillars have brittle irritant hairs, and skunks eject a vile secretion from anal glands in the direction of the attack. In all cases, however, if attack can be prevented by threat of retaliation, as with the raising of hair or feathers, snarling or hissing, so much the better. In some species that have only feeble weapons threat may be complete bluff, but be just as effective.

ME/JE

ANIMAL ARCHITECTS

Homes to live in. . . Sand cases of amoebas. . . Nests of birds. . . Beaver dams and lodges. . . Animal burrows. . . Cocoons of moths and butterflies. . . Traps to capture prey— spiders' webs. . . Signals to other individuals—bowerbirds. . . Weaverbird nests. . . Underground air conditioning of termites and prairie dogs. . .

▶ **Home for stinging hoards**—the elegant papery nest of the social wasps *Polybia occidentalis* hanging below a *Heliconia* leaf in Trinidad.

LONG before they landed, the occupants of an alien spaceship approaching the earth for the first time would probably realize that our planet was populated by creatures of high intelligence. From far above the atmosphere, their scanners would reveal perhaps the most outstanding marks of civilization— buildings, from skyscrapers to small houses and from giant dams to nuclear power stations. But it is not only humans that construct edifices although their scale and influence are usually of a higher order than the constructions of the rest of the animal world. Animal architects have also achieved some remarkable feats of engineering in building beaver dams, tunneling complex badger setts, or weaving intricate birds' nests.

The nest of a hummingbird seems such a remarkable achievement that it is tempting to believe that the builder has some awareness of what it is building and, in fact, birds certainly do have advanced nervous systems. But, even some amoebas (Protozoa) build beautifully intricate sand-grain cases although they are single celled and lack a nervous system. In doing so, amoebas show exactly the same building processes as a bird building a nest. The pseudopodia of the amoebas pick up appropriately sized sand grains as they flow over the bottom

of the pond and store them as a mass inside the cell; the animal must collect enough particles to make a complete new case. When the cell divides, the sand grains come to the surface of the new daughter cell and are arranged so that differently sized particles are in different places and the case has its characteristic architecture. So, a sophisticated nervous system is not essential; talented builders and architects can be found throughout the animal kingdom.

Some groups of animals are great builders and others are not. There are many hundreds of species of birds which build intricate nests of varied design whereas, although there are numerous species of burrowing mammals, few have impressive building skills. Notable exceptions are beavers which build dams that raise the water level of a river to create a large lagoon in which a massive log dwelling or lodge is built.

Looking at the animal kingdom as a whole, there are really only three groups that can be described as outstanding builders: the spiders, the insects and the birds. Body size seems to have a bearing on this. Mammals are comparatively large animals, and only the smaller ones, mainly burrowing rodents, tend to be builders. Small rodents construct tunnels and caverns for themselves and their young to live in. In this way, they are protected from heat, cold, rain and desiccation as well as being safe from predators. An elephant or even a small antelope is large enough not to suffer so seriously from the effects of climate and is more effective at detering predators or running

Underground Air Conditioning

The antipredator defenses of some structures are so formidable that the occupants would be in danger of dying of suffocation within them were it not for special air conditioning systems built into the design.

The most advanced termite species (members of the genus *Macrotermes*) live in mounds (**1**) which reach a height of 5m (16ft) and have mud walls about 50cm (20in) thick surrounding the living area. Vital oxygen to keep the termites and their fungus gardens alive is carried to the nest cavity through a ventilation and air conditioning system. This system varies to some extent from one part of their native Africa to another but, in its most elegant form, it is completely enclosed. The power to drive the system comes from the heat produced in the living area by the metabolism of the termites and their fungi. This causes air to

rise through the chambers of the nest into a spacious attic, from which it is pushed into fine channels which run down ridges in the mound surface. It is here that fresh oxygen

diffuses in and carbon dioxide diffuses out. The temperature of the air in these superficial channels also drops and it is carried down into a cavernous cellar where it becomes humidified by ground water.

Similarly, the North American Black-tailed prairie dog has a ventilation system which carries air through its burrow (**2**). The burrow systems of these prairie dogs emerge at the surface through exits of two different kinds; rather low, round-topped mounds and taller steep-walled craters. Burrow systems usually have one of these exits at each end and, as a consequence of their height and profile difference, the speed of the wind passing over the crater is faster than that passing over the mound. This causes a pressure difference between the two, so drawing air out through the crater and in through the mound.

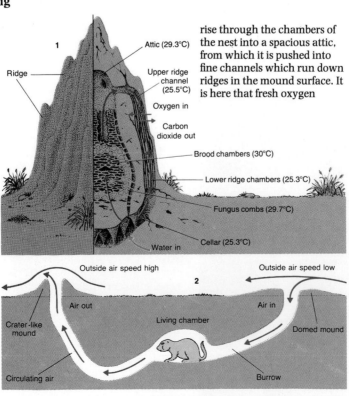

1
Ridge
Attic (29.3°C)
Upper ridge channel (25.5°C)
Oxygen in
Carbon dioxide out
Brood chambers (30°C)
Lower ridge chambers (25.3°C)
Fungus combs (29.7°C)
Water in
Cellar (25.3°C)

Outside air speed high
Outside air speed low
2
Air out
Air in
Crater-like mound
Living chamber
Domed mound
Circulating air
Burrow

away than at building a dwelling. Interestingly, one of the largest animals to build a nest is the Wild boar. Among hoofed animals, it is unusual in that this pig produces a litter of rather small young which need protection from cold and which would have difficulty in escaping from predators such as wolves.

Another feature which characterizes the most accomplished builders is manipulative skill. In birds it results from a combination of sharp vision, a very mobile head and a precision tool—the bill. It is this combination which has enabled some weaverbird species actually to tie knots with strands of grass. In the case of spiders and insects, it is jointed limbs and delicate mouthparts which have provided the manipulative skill. Certain other attributes have also contributed towards the evolution of building ability in these three groups. Taken as a whole, insects are remarkable architects but the greatest achievements are confined largely to only four insect orders: the butterflies and moths, the caddisflies, the ants, bees and wasps, and the "white ants" or termites. The extraordinary elaboration of nests in ants, bees, wasps and termites parallels their social development which far exceeds that of any other insect, indeed, any other invertebrate. In this case, therefore, the evolution of nest architecture has gone hand in hand with the evolution of advanced social organization.

The building ability of butterflies, moths and caddisflies is actually confined to their caterpillars or larvae. Many of these build protective cocoons in which to chrysalize or pupate and, particularly among the caddis larvae, many also build portable larval cases. The feature that characterizes these insects is their

Nest Weavers

One of the nest-building techniques of birds is to twist and loop together strips of leaf to make a roofed hanging basket. Not a drop of adhesive is used, and the structure is held together by the frictional resistance of the material. No single family of birds is more expert at this technique than the weaverbirds. Weaving is a rather misleading term for the often irregular arrangement of strands, but these birds can carry out a variety of distinct "stitches," including interlocking loops, spiral winding and even knots. Such control over the building material could not be achieved with dry plant material; it has been a decisive evolutionary step for weaverbirds to tear strips of grass and palm leaf from living green plants.

In the Village weaverbird, the nest, which is built by the male, starts as a ring attached below a forked twig. He then stands in the ring to attach each successive strip of material, leaning forward to complete the egg chamber. Each new strip of vegetation is added by being pushed into the fabric, pulled through, pushed in again, and so on—much more like needlework than weaving. When the egg chamber is completed, the male builds the entrance porch leaning gradually backwards from his position in the original ring.

This very sophisticated building behavior of weaverbirds does seem to be at least partly learned. The first nest built by a young male Village weaverbird is very rough looking and his technique lacks the skill shown by experienced birds. Sometimes, he does not push in the strip far enough so it falls out, pushes it in but then pulls it out again or pushes it in but fails to thread it up completely, leaving a loose end.

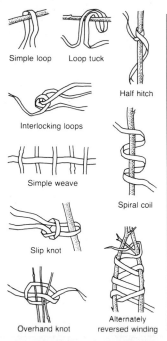

Simple loop   Loop tuck

Half hitch

Interlocking loops

Simple weave

Spiral coil

Slip knot

Overhand knot   Alternately reversed winding

◄ **Waiting to capture prey,** an orb spider (*Argiope* species) sits in the center of its web. The thicker bands, known as stabilmenta, around the middle make the web conspicuous to birds which would break the web if they blundered into it.

► **Weaver stitches.** A range of stitches used by weaverbirds when constructing a nest.

▼ **Pride in its workmanship**—a Baya weaver (*Ploceus philippinus*) puts the final touches to his masterpiece.

ability to secrete silk which they use to spin all-silk constructions, such as the cocoons of silk "worms," or use as a cement to stick together sand grains, sticks or cut leaf panels. Silk filaments, with their great strength coupled with lightness, have been as crucial in the evolution of case and cocoon building in these insects as in the evolution of prey capture devices by spiders.

The most important single function of animal buildings is as a place to live. Sometimes, these are for the use of the builder alone, as are the portable cases of caddis larvae, while others, such as birds' nests, are built especially to raise young. Still others, such as the nests of ants or termites, are homes for whole families or colonies. These houses incorporate design features which protect the occupants from the hazards of life, climate and predators. The caterpillar of one species of Central American euchromid moth (*Aethria* species) is covered in tufts of long hairs to protect it against bird and insect predators. This defense would, of course, be lost when the larva molted its skin to become a pupa except for the fact that the caterpillar first plucks out its hairs one by one and arranges them in whorls around the plant stem above and below itself before finally incorporating the remaining hairs into the silken cocoon inside which it pupates.

The second most common function of animal construction is to capture prey. The majority of examples of this come from the silken traps of spiders and the best known of these are the species which build orb-shaped webs. Remarkably, identical orb web designs appear to have evolved in two quite separate groups of spiders whose capture threads work in completely different ways. In one group, which includes some rather large spiders with webs 60cm (2ft) or more wide, the capture threads are covered in sticky droplets like a row of tiny beads. In the other group, which are generally rather small spiders, the capture thread is dry but covered with a fuzz of fine threads with loops and barbs that snare the prey. Some species of spiders have webs that are clearly specialized for the capture of particular types of prey. A special problem for spiders is the capture of butterflies and moths; these have loosely attached scales covering their bodies which may allow the insect to pull itself free from the web with the loss of only a few scales. One spider, which specializes in moth capture, builds a horizontal triangular web with only a few capture strands hanging loosely between a central and a frame thread. These capture threads have very large sticky droplets which overwhelm the protection that the moth gets from its loose scales. As the moth tries to jerk free, the capture thread breaks at a special point of weakness at its outside end, leaving the moth whirling round at the end of its tether until the spider kills it.

Finally, there is another function of animal constructions which deserves mention—use as social signals. For example, males of a species of land crab, the Ghost crab, make sand pyramids outside their burrows, which attract females and deter males. Most elaborate of this type of construction are the courtship arenas of bowerbirds. One of the most remarkable of these birds is the Brown gardener, which is dowdy in appearance, as the name suggests. The male builds a spectacular tent of twigs with a courtyard decorated with little piles of berries and flowers which may be changed every day.   MHH

RHYTHMS OF BEHAVIOR

Types of rhythm. . . Responses to light levels. . . Control by the environment or an internal clock. . . Annual breeding cycles. . . Sleep. . . Hibernation. . .

ALL animals live in an environment in which some of the events affecting their lives are rhythmic. For example, animals may need to change their behavior at or before dawn, high tide, full moon, or the onset of warm spring weather. Behavior rhythms may result from responses to environmental rhythms or they may be a consequence of internal rhythms.

The term "rhythm" refers to a series of events repeated approximately regularly, but a truly regular rhythm, in which the events are separated by equal periods, is said to be a "periodicity." The period between events or wavelength of the rhythm can range from fractions of a second to years. The wingbeat rhythm of the smallest insects has a period of one-thousandth of a second while, for many eye-tracking, breathing, chewing or walking movements, the period is less than five seconds. Rhythms of feeding and digestion have wavelengths of minutes or hours, according to the size of the animal, while activities which depend on the state of the tide or the light level will have wavelengths of 12.5 and 24 hours respectively. Longer-wavelength rhythms include the reproductive cycles of female mammals and responses to the phases of the moon.

It is apparent from this wide range of behavioral rhythms that some are merely a consequence of control mechanisms but others serve a useful function for the animal, allowing it to cope with its environment. Species which can find food, avoid predators and generally maintain body state more effectively during daylight than during darkness obviously benefit from responding appropriately to different light levels and may operate most efficiently if they can predict the changes during the 24-hour period by means of an internal clock. The same argument would apply to responses to any other environmental rhythm, whether physical or resulting from the activities of other animals. Fishes in a school and birds in flocks may benefit from assuming a rhythm of locomotion which is precisely the same as that of their fellows and, if the main predator that might eat an animal comes near every two hours, then it is obviously advantageous to synchronize antipredator behavior with this rhythm. Horseflies and mosquitos may be able to bite mammals most effectively during the dawn and dusk periods. The activity peaks of parasites may be related to host behavior; for example, the human threadworm female often moves down the gut to the anal region to lay her eggs in the evening when the host is most likely to scratch and increase the chance that the eggs will be ingested. Avoiding predators may be more effective if the activity of the group is synchronized at a particular time of day. Newly fledged Brünnich's guillemots, for example, are vulnerable to attack by large gulls but most of the chicks leave their cliff nests within a very short time span so many can survive the depredations of the limited number of predators.

Breeding by large animals often occurs annually in temperate regions. It can be initiated by a particular day length. The interval between breeding seasons is not always one year in

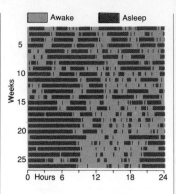

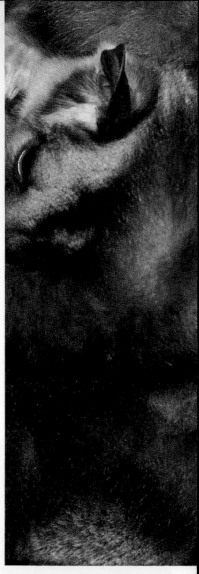

▲ **Sleep patterns in a human baby.** Circadian (24 hour) cycles of sleep and wakefulness of human babies become established 16 weeks after birth.

▶ **Daily ritual**—Sugar gliders (*Petaurus breviceps*) settle down for their daytime sleep in a communal nest within a tree in the Australian woodlands. They are active at night.

▼ **Spawning frenzy.** Grunions (*Leuresthes tenuis*) of California respond to phases of the moon by coming to deposit eggs high on the beach during spring tides. The young develop well enough to be freed from the eggs by the next spring tide.

the tropics, however, for seabirds, such as some shearwaters and boobies, breed at intervals of eight to nine months. Breeding is annual in some invertebrates, such as the Palolo worm of the Atlantic and Pacific which gathers in vast numbers for fertilization during the neap tides of the last quarter moon in October and November.

During periods of adverse conditions in which finding food is difficult, or there is a high risk of predation or physiological damage, animals may reduce their activity and their metabolic rate. If the adverse period is the night, or the day for nocturnal animals, the usual behavioral response is sleep. Sleep is a prolonged period of inactivity with reduced responsiveness and a characteristic posture. The site chosen is usually appropriate for avoiding predators and the behavior often occurs with a circadian (approximately 24-hour) or tidal periodocity. Sleep may have no restorative function so it seems most likely that it has evolved as a means of minimizing predation risk and conserving energy. It does form part of the daily rhythm, however, and individuals may be harmed if that rhythm is disrupted by sleep deprivation.

Where sleep is associated with reduced metabolic rate and some lowering of body temperature, the word "torpor" is used. Torpor lasting for many days or months is called "hibernation" in winter conditions and "aestivation" in adverse summer conditions such as long dry periods in deserts. Torpor is the normal nocturnal state for cold-blooded animals in temperate regions, and the bumblebee buzzing to warm its body on a cold morning is a common sight. Bats and other small mammals as well as birds, such as hummingbirds, can become torpid overnight and hence save energy. Larger animals need too much energy to raise their body temperatures from low levels for overnight torpor to be worthwhile.

Before hibernating an animal must feed more to accumulate fat deposits as an energy store, select a suitable site and, often, build a nest. During hibernation, small mammals may lower their body temperature to a few degrees above ambient with seventy-fold energy savings, and may awaken on cold days. Mammals weighing 1 to 5kg (2 to 11lb), such as hedgehogs and marmots, do not awaken briefly because of the high energy cost and the long awakening time. A small bat can wake in half an hour but a marmot needs many hours. For a large mammal, such as a bear, the body temperature can be lowered by only 5°C (9°F), the metabolic rate reduced by only 50 percent and awakening is rapid so that this is scarcely hibernation.

Experiments with constant environmental conditions have shown that many of the periodicities found can be controlled by internal processes only. Recent studies of these have shown that the period can sometimes be lengthened by cooling the animal or by substituting the heavy isotope, deuterium, for normal hydrogen in the water so making the clock run slow. Evidence for the localization of pacemakers in the brain comes from studies in which removal of the pineal in sparrows, or the suprachiasmatic nucleus in rats, results in the loss of rhythms shown in constant conditions.   DMB

ORIENTATION AND MIGRATION

Finding the way around territories. . . Remembering landmarks in digger wasps. . . Sun and magnetic compasses in animals. . . Route reversal strategy in insects. . . Why do animals migrate ?. . . Strategies of migration. . . Eruptive movements of birds. . . Fuel for migration. . . The timing of migration. . . Navigation during migration. . . Magnetic, celestial and solar cues. . . Migration record breakers— locusts, terns, warblers, eels, whales. . . Migration patterns of birds of prey. . .

THE way in which animals are able to orientate themselves is often considered in connection with migration. A small bird, such as a Willow warbler, breeding in central England, will leave in the fall and return in the following spring to the same spot where it bred the year before. It will have travelled around more than a quarter of the globe to avoid the unfavorable conditions of the British winter—obviously this bird needs some sophisticated orientation mechanisms to accomplish its journey and its return to the same place. A closer look at what is known about orientation reveals, however, that many animals which do not migrate—ants and bees, fishes and frogs, sedentary birds and small mammals—have well-developed, highly complex methods of orientation. It becomes clear that orientation capabilities have evolved not only for migration, but primarily to enable animals to move quickly and efficiently from place to place within their home range.

The simplest method of orientation within a familiar area appears to be learning the lie of the land around the home. A classic experiment by Niko Tinbergen, demonstrated the importance of memorized landmarks for the orientation of digger wasps (*Philanthus triangulus*). While the female was inside its nest, a circle of pine cones was placed around the nest entrance. Leaving the nest, the wasps circle around the site memorizing the surrounding structures, and then leave for the next foraging flight. If the cones were displaced a short distance sideways, the returning wasp began to search for its nest entrance at their new position, inside the center of the circle of cones rather than where the hole really was. Memorizing structures surrounding the home in this way need not be restricted to visual landmarks. Any other characteristics of the area which are perceptible to the animal might be incorporated in this picture; for example, odors play an important role in the orientation of many animals.

The usefulness of an animal's knowledge of its home area depends on the number of landmarks it can include in the pic-ture, and their number increases dramatically as the radius of the familar area increases. It is not surprising, therefore, that other orientation mechanisms, or "compasses," have been developed to cope with orientation even within a limited home range.

A compass indicates the animal's geographic directions, that is, the equivalents of our ideas of north, south, east and west. Two such systems which have been shown to play a role in animal orientation are: a magnetic compass and a sun compass. The magnetic compass has been demonstrated in a number of mainly vertebrate animals, and it appears to be a rather simple mechanism, based on the animal's ability to perceive the earth's magnetic field and thereby work out directions. Sun compass orientation has been extensively studied in many arthropods (crustaceans, spiders, insects) as well as in fishes, amphibians and birds. It requires more sophisticated mechanisms involving an internal clock, for the animal has to compensate for the changing direction of the sun in the course of the day. Experiments with honeybees and with pigeons have shown that this compensation is based on experience. The animals have to observe the sun's arc and learn to associate sun position with time of the day and with geographic direction to be able to use the sun for direction finding. Young pigeons that had seen only the descending part of the sun's movement in the afternoon were found to be unable to use the sun compass in the morning; instead they relied on their magnetic compass at this time of the day.

Using an internal compass to localize geographic directions offers new possibilities of orientation. The animal can remember in which direction it leaves home and obtains the course home by returning on a compass bearing which reverses this direction. Such a strategy, known as "route reversal," is applied by many arthropods, such as bees and ants, and it also seems to play an important role in the orientation of young,

▼ **A classic experiment**—the Tinbergen orientation and displacement experiment with digger wasps. The entrance to the underground nest was surrounded by cones. (1) When a wasp leaves the nest it flies around the area presumably memorizing the key features. (2) Before it returns the cones are moved, and on its return it goes directly to the center of the circle of cones rather than to its nest.

▶ **The greatest wanderer** on earth, the Arctic tern (*Sterna paradisaea*) migrates from Arctic to Antarctic and back again, a distance of 36,000km (22,000 miles) each year.

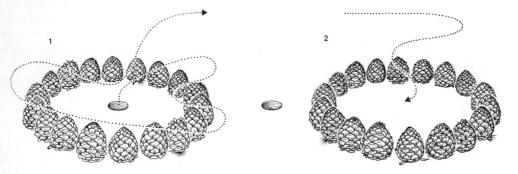

inexperienced homing pigeons. An even more effective strategy will arise when compass orientation is combined with the picture of familiar landmarks mentioned above. The animal forms a directionally oriented "map" of the landmarks within its home range. The path to a desired goal is no longer a chain of familiar places, but, by remembering the courses from characteristic marks to home and to one another, the animal can obtain the desired direction as a compass course. This would allow fast and efficient movements between nest, foraging sites and so on, and it would relieve the animal of paying attention to the outward journey.

The image of the distribution of familiar marks within a home range, or "mosaic map" is important in the navigation of homing pigeons close to home. An extension of this "map,"

to include factors, which are like gradients spreading outwards, would allow birds to extrapolate to give them a "navigational map" by which they could derive their home direction from local factors even in unknown territory. Before sedentary animals began to migrate, it is likely that they possessed well-developed capabilities for orientation within their home ranges, including some type of map and, most important, compass mechanisms.

Migration has evolved among many animal groups. Flying and swimming animals are the record-breaking migrants because of their speed of motion but there are very impressive movements of small land animals, such as newts and frogs, as well as conspicuous migrations of mammals, such as the migrating herds of buffalos in the old American West or of zebras, wildebeest and antelopes in the African savanna.

Animals migrate for many reasons but, essentially, they do so to avoid unfavorable conditions or to look for regions, which offer specific, more favorable conditions. Overpopulation, causing shortages of vital resources, is one of the reasons for mass migration, often ending as journeys of no return. The most spectacular examples are the movements of lemmings and the flights of huge flocks of locusts which destroy crops wherever they feed. Some of the sudden (eruptive) movements of certain bird species, such as the Siberian jays, nutcrackers and waxwings, are also caused by overpopulation. It is not known to what extent these birds use a capacity to orientate to direct their journeys. Although some individuals may return to their traditional breeding grounds, an advantage of this emigration is that the animals may reach new regions which are favorable enough for them to extend the species' distribution range. Irregular movements also occur in many species on the Australian continent. Unpredictable climatic conditions with

Migration Record Breakers

The distances travelled by the various animal species cannot easily be compared. Is, for example, the 500km (300mi) tiresome journey of the crab, *Ericheir sinensis*, from the Elbe estuary up to Prague more of a feat than an occasional airborne ride of a tiny spider on its own silk web over 200km (125mi)? Yet the distance records of invertebrates are definitely held by the insects because their ability to fly allows them to cover long distances easily. Gregarious desert locusts such as *Schistocerca gregaria*, are reported to travel more than 5,000km (3,000mi) from the Arabian peninsula to Mauritania in less than two months. Also, some butterflies migrate regularly; the Monarch butterfly leaves Canada in late summer to travel to Mexico.

Among the vertebrates, the champions of long-distance migration are animals which travel in water or air, that is, fishes, whales and birds. The larvae of the European eel take three years to migrate from the Sargasso Sea to European rivers, and some species of salmon travel for feeding purposes into the open Atlantic or Indopacific Oceans before

they return years later for spawning.

The longest distances, however, are covered by migrating birds. One of the most spectacular journeys known is undertaken by the Arctic tern, which breeds along the arctic coasts. The population of northern Canada migrates southeastwards, crosses the North Atlantic and then flies southwards along the European-African coasts. While many birds winter in South Africa or cross the Atlantic again and winter along the coast of Tierra del Fuego, others continue their trip until they

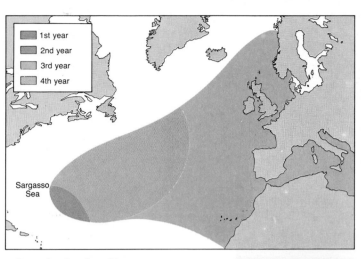

▲ **Stages in migration** of European eel larvae (leptocephali) from the Sargasso Sea where they hatch in March of Year 1. Through the following years they migrate across the Atlantic reaching the Western European seaboard in Year 3 and the Mediterranean in Year 4.

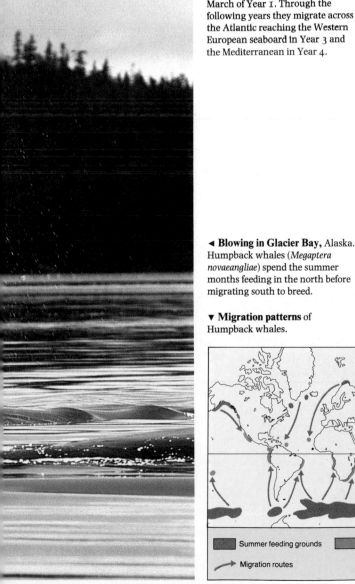

◄ **Blowing in Glacier Bay,** Alaska. Humpback whales (*Megaptera novaeangliae*) spend the summer months feeding in the north before migrating south to breed.

▼ **Migration patterns** of Humpback whales.

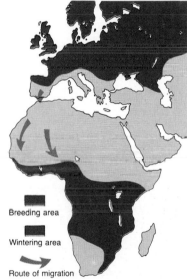

▲ **Migratory routes** of Garden warblers. Note that birds from central Europe migrate through Spain.

reach the antarctic shore, and they may circle the antarctic continent before they start the return flight to their northern breeding areas. The transoceanic journey of the west-Alaskan population of the Bar-tailed godwit, that spends the northern winter at the coasts of Australia and New Zealand, is equally impressive. But it is not only birds that feed along the coasts or are ecologically connected with the sea which travel extremely long distances. Many small passerines, such as the Willow warbler, with a body weight of only about 7g (0.25oz) that migrates from northern Scandinavia to southern Africa, cover thousands of kilometers twice a year. Such birds migrate mostly over land and are able to find refueling areas on the way. In North America, however, some species, including the Blackpoll warbler, are reported to leave the coast at Cape Cod and reach the southern Caribbean islands and the coast of South America after three or four days of travel over the open Atlantic ocean; for this flight, they take advantage of weather conditions which provide strong tail winds during most of the journey.

▼ **Migration routes** of Arctic terns that breed in northern Canada. Birds winter off South Africa, off southern South America, or circle off the Antarctic, before returning along their original tracks to Canada.

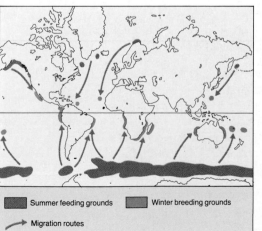

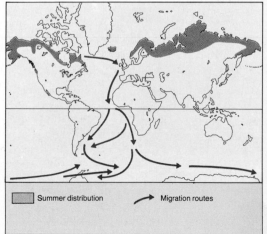

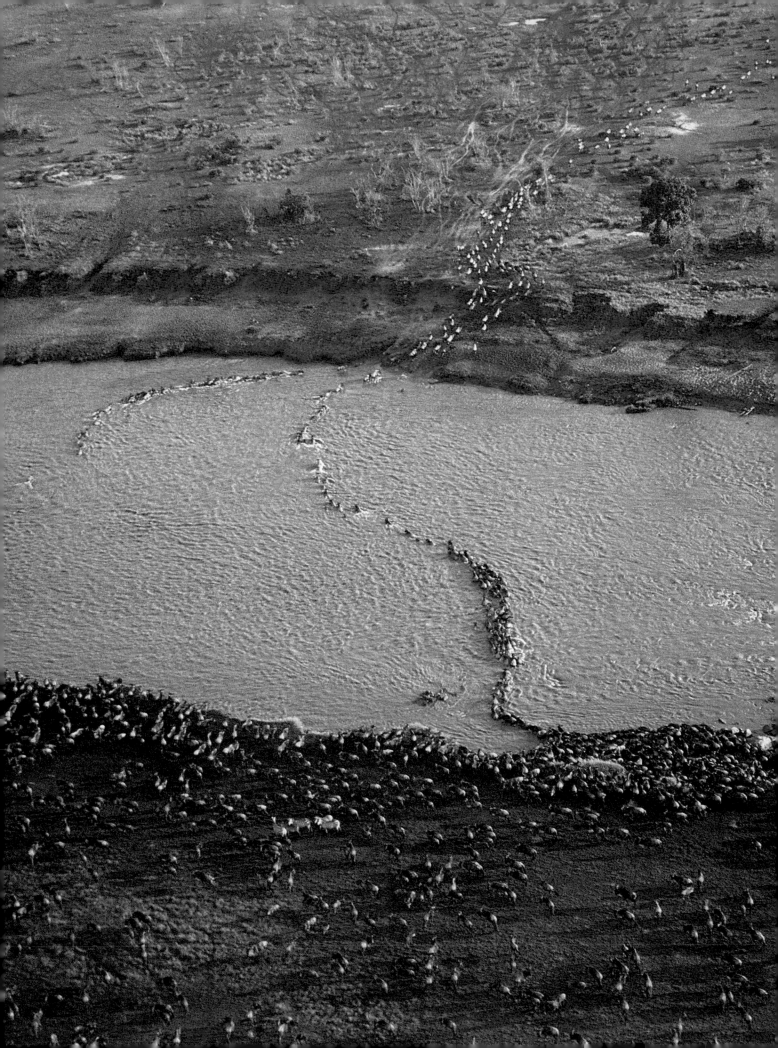

rare, sporadic rains force desert birds, such as Zebra finches, budgerigars and so on, to live a nomadic life until, by chance, they find a region where rainfall triggers a new period of vegetation growth. Here they become temporarily sedentary and breed.

Migratory movements are often associated with reproduction. In many animal species, the development of eggs and young requires specific environmental conditions which are not available where the adults normally live and feed. Familiar examples are the spawning migrations of terrestrial amphibians which, each spring, lead frogs, salamanders and so on back to ponds and streams. Whales, feeding in the cold arctic or antarctic waters, migrate to the warm equatorial seas to give birth to their calves. Many small fishes, feeding on the plankton in the warmer water of lakes, must migrate into the tributary streams for spawning because their eggs and larvae need a lot of oxygen. The spectacular migrations of species of salmon occur for similar reasons. Their offspring develop in highly oxygenated freshwater, while the adults feed in the open ocean until, years later when they have become mature, they return to their stream of origin for spawning.

The migration of eels remains somewhat mysterious. Here the adults leave the ocean to feed in freshwater and, at the end of their lives, they migrate back to the Sargasso Sea to reproduce.

The most widespread migratory movements are brought about by seasonal climatic changes. Animals migrate to avoid periodically adverse conditions or to exploit food supplies which are only accessible at certain times of the year. Many mammals, for example, spend the summer in the mountains while they descend to the valleys in winter. In regions with rainy and dry seasons, animals look for the areas which offer the best food supply. Most prominent, however, are the annual migrations of many birds and some species of butterflies that live in higher latitudes and migrate to escape the harsh winter conditions of these regions.

The extent of a species' migration is often determined by its food requirements. For example, purely insectivorous birds may be forced to leave a region, even if the average winter temperatures are above freezing, because food is not available, while seed-eating species normally migrate shorter distances and spend the winter farther north. In severe winters with heavy snowfall, however, they may extend their migration considerably. Besides the availability of the food, daylength, that is the time during which a bird which is active in the day can spend foraging, becomes an important factor and forces birds living at high latitudes to move south.

Different flying and feeding habits have led to the development of various strategies of migration. Large soaring birds migrate strictly during the day taking advantage of thermal updrafts. For this reason they prefer to migrate over land, and they frequently concentrate in huge numbers along mountain ridges with regular updrafts or at isthmuses. Famous spots to

◄ **River crossing**—no barriers are big enough to prevent Blue wildebeest (*Connochaetes taurinus*) in their annual migration to find new sources of food.

observe hawk migration are, for example, Hawk Mountain, Pennsylvania, USA, and the capes of Tarifa and Gibraltar in southern Spain. Birds which feed on airborne insects, such as swallows and swifts, are day migrants, too, feeding while they migrate. Most birds flying in flocks, such as many species of waterfowl and several seed-eating species of finches, also migrate during the day. The majority of small passerine birds, however, migrate at night. The main advantage of this strategy is that, at a time of the year when the period of darkness is relatively long, these birds can spend the hours of daylight searching for food in the unfamiliar places where they rest during migration. In addition, the lower night temperatures and the usually calmer winds are more favorable for the migratory flight.

Some bird species cover extremely long distances during migration, and it is important to understand how the birds find enough energy to fly so far, especially when crossing seas or deserts, that is, regions where a regular intake of food is impossible. The physiology of birds is adapted to master these problems: before they start migrating, most bird species build up fat deposits which serve as "fuel" for their migratory activity. Some small passerines double their body weight before migration and, in every case, the amount of fat deposited correlates well with the distance to be travelled. Fat has the advantage of yielding more energy per gram than any other body substance. Unlike mammals, birds can convert it directly into energy, carbon dioxide and water; water gained in this way might temporarily be the only source of moisture for migrants while crossing adverse terrain.

The timing of migration is a complex phenomenon involving the interaction of external and internal factors. Behavioral rhythms (see p44) are synchronized with the seasons by changing daylength. Together, they control migratory behavior and cause the birds to leave at the appropriate time. The distance of migration is also, in part, controlled by this internal calendar which, by controlling the duration of migration, determines the distance travelled. This brings a bird into the wintering area of its species. Here, the migration need not end abruptly, but the bird can search around until it has found a suitable habitat where it will spend the winter.

The way in which birds navigate during migration is a

◀ **"Seeing" in the dark.** Living in dark environments or being strictly nocturnal forces animals to evolve eyes of extreme sensitivity at low light levels or to sharpen other senses, using, for example, echolocation. Here the animal itself emits sound and the returning echoes are analyzed to obtain information about distance, size, shape and motion of objects in the animal's environment. Echolocation of low efficiency is used by two species of cave-nesting birds including the oilbird (*Steatornis caripensis*) LEFT, fruit bats and some species of rodents and insectivorous mammals. Only whales and bats of the suborder Microchiroptera have developed highly sophisticated, very efficient sonar systems using frequencies too high for us to hear.

▶ **Transparent elvers** after their long migration from the Sargasso Sea and ready to swim up European rivers where they develop into adult eels.

▼ **Displacement experiments with starlings.** Birds migrating in a direction slightly south of due west were captured in the Hague, Holland, and released in Switzerland. (1) Juvenile birds that had never migrated before continued to fly in the previous direction and thus arrived in the wrong wintering area. (2) Adult birds adjusted their subsequent course to north-west so that they arrived in their traditional wintering grounds.

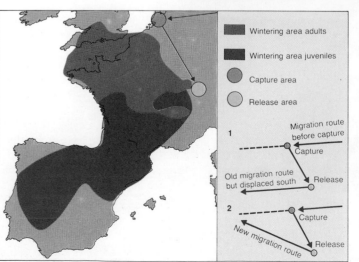

Wintering area adults

Wintering area juveniles

Capture area

Release area

1   Migration route before capture

Capture

Old migration route but displaced south   Release

2   Capture

New migration route   Release

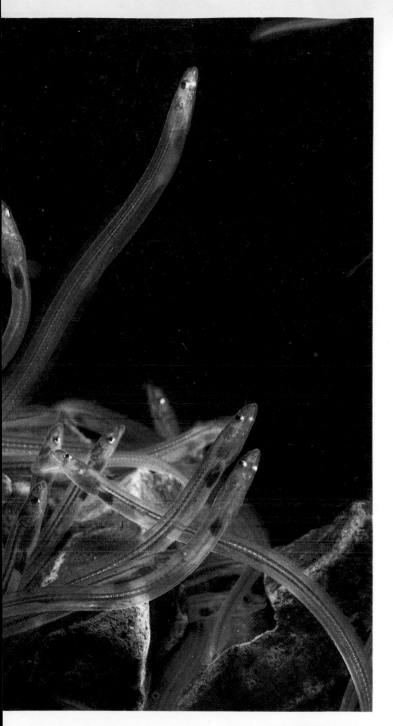

The inbuilt direction-finding mechanism requires some external reference. Orientation mechanisms had already developed in sedentary birds and hence were available when a species began to migrate. Thus, the only step necessary to enable birds to navigate on migration was to couple genetic information about a set direction to an already existing compass. It has been found that many bird species use a magnetic "compass" to set their migratory direction. Laboratory experiments with handraised birds that had been isolated from celestial cues indicate that the magnetic field provides a reference system for innate directional information which is independent of other cues, and sufficient to guide the young birds. This directional information need not be constant but may be modified by the rhythm which controls the duration of migration. For a young Garden warbler, born in central Europe and migrating via Iberia to central Africa, the migration pattern may be something like: "Fly six weeks southwest, then eight weeks southsoutheast."

As well as the magnetic field, many nocturnal migrants use celestial information. The positions of stars vary during the night as well as during the course of the year. The birds master these problems by using the constant spatial relationship between star patterns to derive their migratory direction, just as we are able to find "north" from the constellation of Ursa Major, regardless of its current position. Experiments with hand-reared Indigo buntings showed the ability to use stars was learnt and that celestial rotation was the underlying reference: "south" may be recognized as "away from the center of rotation." Such a star compass functions independently of the magnetic compass and represents a unique development of night-migrating birds.

The migratory direction may be represented in relation to the magnetic field and to the center of celestial rotation in the same bird species. At first, both systems appear to be independent; during actual migration, however, directional information from the magnetic field affects the star compass which may be recalibrated. Experimental evidence indicates a highly complex interrelationship between magnetic and celestial information, which is not yet completely understood.

The sun compass has been frequently suggested as an important mechanism for daytime migrants, but experiments suggest that it probably does not play a major role. The reason for this might be the dramatic changes of the sun's arc with geographic latitude. Many nocturnal migrants, however, seem to use the setting sun, not as part of a sun compass, but as a pointer indicating west. It is not clear yet whether the sunset point represents another independent source of information or whether it is calibrated by other systems; recent experiments seem to suggest the latter.

On a young bird's first migration orientation cannot be based on experience but any later migration means returning to a location where the bird has stayed before. Thus, older migrants can head for their goals—breeding place, winter quarters or refueling areas between—using mechanisms based on a navigational "map," at least after they have reached the general area of their destination. Their orientation may now be very similar to that of homing pigeons returning after a long displacement.   ww

fascinating subject. Every year, many young birds face the problem of moving to their winter quarters which are, at that time, still unknown to them. Some species migrate in flocks; here, experienced adults may guide the young, and the migration routes and wintering areas are passed on, at least in part, by tradition. But many species migrate singly and the individuals must have information of their own telling them where to go. Large-scale displacement experiments have revealed basic differences between adult birds and young ones migrating for the first time. While adults compensated for the displacement and returned to their traditional winter quarters, the young birds maintained their course reaching a new, parallel-displaced wintering area. Adult birds are familiar with their destination and were able to head for it. Young birds, however, still lack this knowledge; they rely on information for the distance and direction of their first journey which is genetically based.

The Long Haul

Migration patterns of birds of prey

Watching raptors on migration has become a favorite pastime for thousands of bird watchers. This is possible because certain species have well-defined migration routes, with concentration points, over which at migration times continual streams of birds pass within the visual range.

Nearly every raptor species that has been studied performs some sort of migratory movement in at least part of its range. The longest journeys are made by those birds which fly regularly between eastern Siberia and southern Africa (eg Eastern red-footed falcons), or between northern North America and southern South America (eg tundra-breeding Peregrine falcons). Even on the shortest routes, this entails some individuals flying more than 30,000km (18,650mi) on migration each year. Journeys almost as long are made by Western red-footed falcons, Lesser kestrels and Steppe buzzards that travel between western Siberia and southern Africa, and by the Swainson's hawks that travel between Alaska and Argentina. Long sea crossings are made by other species, for example by the Eastern red-footed falcons which cross from India to southern Africa, and by the Lesser sparrowhawk which migrates between Japan and the East Indies. Long desert crossings are made by the many species that travel between Eurasia and tropical Africa.

All raptors migrate by day but, while some species progress mainly by active flapping flight, others progress mainly by soaring, climbing in successive updrafts or thermals and gliding with some loss of height to the next. The first group includes the narrow-winged raptors, such as falcons and harriers. They expend considerable energy on migration, but can cross water with ease, and can thus migrate on a direct course. Their movement proceeds on a broad front, and at no point on the route do really large concentrations of birds occur. The second (soaring) group includes the broad-winged raptors, such as eagles and buzzards. They expend less energy on migration but, being dependent on updrafts, they are obliged to make as much of their journey as possible overland, on what is often an indirect route, taking advantage of land bridges or short sea-crossings. In the absence of mountain updrafts, they are also restricted to the middle part of the day, when thermals are best developed. In an intermediate category are the goshawks and sparrowhawks, which take advantage of updrafts when available, but rely less heavily on them than do other broad-winged species. The latter provide some of the best examples in birds of narrowly channelled migration routes. On their journeys between Europe and Africa, thousands of soaring raptors circumvent the Mediterranean Sea each year via the Straits of Gibraltar in the west, or the Bosporus and Dardanelles in the east. At these points the birds form concentrated migration streams, to the delight of bird watchers.

Another more eastern route for broad-winged raptors into Africa lies between the Caspian and the Black seas, joining the Bosporus stream in the Levant and crossing near Suez. Very many birds take this route, and recently some 380,000 raptors of 28 species were counted at the eastern end of the Black Sea between 17 August and 10 October, including 205,000 Steppe

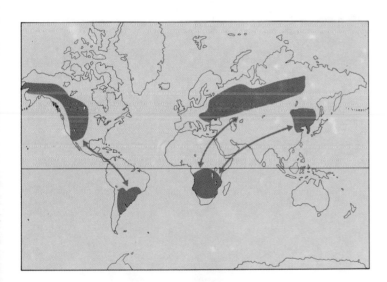

◄▲ **Two migratory birds of prey.**
LEFT Swainson's hawk (*Buteo
swainsonii*). ABOVE Red-footed falcon
(*Falco vespertinus*).

■ Swainson's hawk

■ Red-footed falcon

▲ **Migration routes** of Swainson's
hawk and the Red-footed falcon.
Note that both species avoid long
sea crossings.

buzzards and 138,000 Honey buzzards. This compares with
maximum totals recorded during several falls at Gibraltar and
the Bosporus of about 194,000 and 77,000 raptors of all
species. A fourth route is probably from Asia across Arabia,
and into Africa over the south end of the Red Sea, but it is
not known how many birds are involved. Fifthly, a small passage occurs down Italy, over Sicily and Malta, and across to
North Africa. On the return migration in spring, no less than
605,000 raptors of about 30 species have been counted at the
northern end of the Red Sea in Israel. Further north this stream
splits into several smaller ones, including the Bosporus and
Black Sea contingents.

Less marked migration routes occur in North America
where, in contrast to Europe, the main mountain chains run
north–south. Cross-winds hitting the ridges create updrafts
which are good for soaring, so major migration streams occur
along the Rockies in the west and along the Appalachians in
the east. They also occur around the edges of the Great Lakes
and along various coastal peninsulas, as at Cap May in New
Jersey. Particularly well known as an observation post is Hawk
Mountain in the Appalachian chain of Pennsylvania, where
10,000–20,000 raptors of 15 species have been counted during the fall in most of the last 40 years. For the journey into
South America, the main routes are down the Florida Peninsula across the West Indies and on to the mainland, and also
through Central America over the land bridge at Panama. The
main species on this last route are the Broad-winged and
Swainson's hawks, whose numbers again run to hundreds of
thousands.

Interest in recent years has centered on whether annual
counts at concentration points can be used to give long-term
population trends. The birds are drawn from wide areas, but
it is not known what proportion of individuals take these routes
each year, and what proportion of those which pass the concentration points are seen and counted. If these proportions
vary greatly from year to year, the counts are of limited value.

IN

NAVIGATION AND HOMING

Navigational ability in the Manx shearwater... Laysan albatrosses... Odor and salmon migration... Maps and compasses... Using the sun and the earth's magnetic field... Pigeon homing...

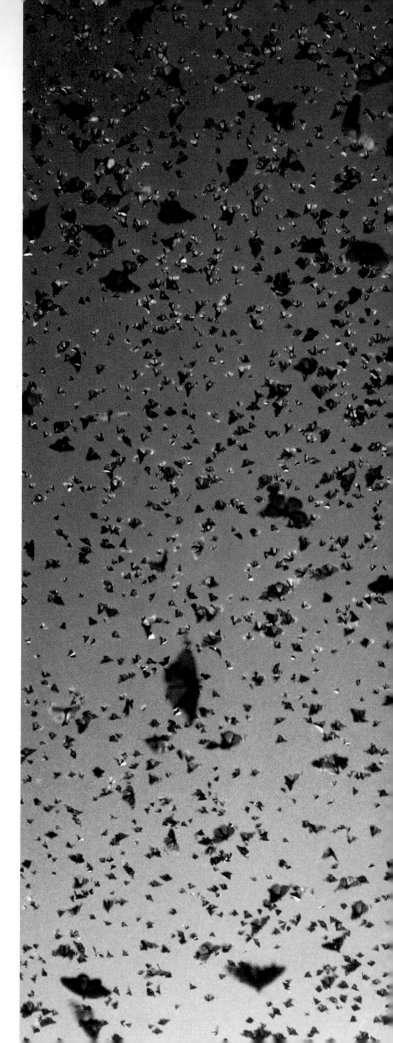

O N a fall afternoon, an endless succession of Monarch butterflies passes an observer in the eastern United States. At night, the skies are filled with migrating birds heading for their wintering grounds. And some species of salmon and trout begin their journeys from the open ocean to the streams where they spawn. All these movements are migrations.

An example of extraordinary navigational ability was the Manx shearwater, taken from its nesting burrow off the coast of Wales and flown by jet to Boston, Massachusetts. Released there, it flew back to its burrow, where it was found 12 days later. Of seven Leach's petrels released in England, two returned to their nests on Kent Island off the coast of Maine. This trip of over 4,800km (3,000mi) was made in less than two weeks! Even more spectacular was the return of Laysan albatrosses to their nests on Midway Island in the mid-Pacific from distances of 4,800–6,500km (3–4,000mi)! These are extraordinary feats which place enormous demands on an animal; not only does it have to fly long distances, but it also has to arrive at a specific place—Midway Island is a tiny target to aim at from 6,500km (4,000mi) away!

The obvious question is, what information do these animals use to find their way? We know that, for homing salmon, odor is an important cue. If we capture a salmon swimming up a stream and return it to a lower part of the river, below a fork, it will choose the correct fork almost 100 percent of the time. But, if we block its nostrils with wax, it chooses the two forks randomly. A particularly dramatic example occurred when a flood washed the salmon out of a hatching pond and into a local river. When these salmon returned to spawn several years later, they were observed flopping along in the almost dry overflow ditch leading from the hatchery down to the river!

Yet, there must be more to a salmon's homing than odor. How could a salmon in the middle of the Pacific Ocean find the mouth of its home river? It is hard to believe that the oceanic phase of salmon homing also depends on odors. At present, we have no idea how they do it, but what seems certain is that they, and other animals that navigate, must have more than just a simple compass.

For an animal moving between its winter home in South America and its breeding ground in North America, we could imagine that a compass might be sufficient. All the bird has to do is fly north in the spring and south in the fall. Yet, many birds return to the exact location where they nested or wintered the previous year, despite crosswinds, storms or other interference with the trip, so that they must be using something more than a simple compass. This is particularly true of birds, such as the Laysan albatross, which are displaced from their homes. No compass alone will be useful unless they can somehow find the direction in which to home. It is this process of knowing the direction in which to home that is called true navigation or the "map." And, because many animals can return

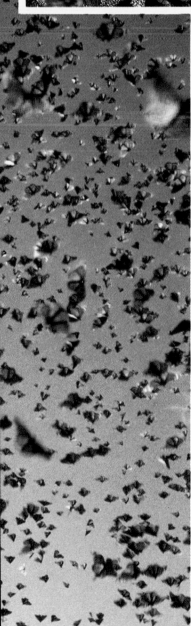

◄▲ **Butterflies on the wing.**
Monarch butterflies are giants
among insect migrants. They leave
Canada in late summer to travel to
Mexico where they winter in
spectacular numbers ABOVE in a few
valleys of the Sierra Madre.

▼ **Demonstrating the earth's
magnetic field.** The length of the
arrows depicts the strength of the
field and the direction of the arrows
the dip and polarity (direction) of
the field—the arrow head indicates
northseeking end of the compass
needle. See text right for further
details.

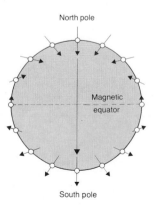

North pole

Magnetic
equator

South pole

home, it can be said that they are clearly able to navigate, to
find their position or they have a "map." In theory, a "map"
alone would be enough to get an animal home. But, it usually
appears that the "map" gives an animal a direction in which
to travel and that animals use a compass system to find and
then to fly, walk or swim in that direction. An animal's "map"
may simply consist of the direction to move—somehow knowing that home is northwest.

The basis of this "map" is almost totally obscure. Human
navigators have long used the sun to determine their position.
By comparing the sun's position in the sky with its remembered
position at home at the same time of day, it is possible to locate
accurately where you are. If, for example, the sun rises two
hours later than it does at home, then the location is to the
west of home. If, at noon, the sun is higher in the sky than
at home, then, in the northern hemisphere, the position is
south of home. Somewhat similar reasoning can be applied to
star patterns. Yet, in no case is there any evidence that animals
use this potential source of information. At present, the nature
of this "map" remains as mysterious as ever.

It has often been suggested that animals might use a
magnetic compass.

No-one really knows exactly how the earth's magnetic field
is generated but, in this case, it can be likened to the result
of pushing a giant bar magnet through the center of the earth
with the ends of the magnet at the magnetic north and south
poles. To explore the magnetic field that results, consider using
a compass (see diagram left) in which the needle is free to swing
in any direction as well as the horizontal plane. At the north
magnetic pole, the north-seeking end of the compass needle
would point straight down. At the south pole, the needle would
reverse and the north-seeking end of the needle would point
straight up. At the magnetic equator, roughly half-way
between the north and south magnetic poles, the needle would
be horizontal with its ends pointing at the poles. But, between
the pole and the equator, the needle would point down at an
increasingly steep angle as the pole is approached. In addition
to this change in the angle, the strength of the field also
changes. The field is strongest at the poles and weakest at the
equator. This means that an animal that could measure either
the angle of the field or its strength would have some indication
of where it was on a north-south line. It would also be able
to use the direction of the field as a compass, except at the north
or south poles, where the field is vertical.

The magnetic north pole and the magnetic south pole are
not located at the geographic poles. The bar magnet through
the earth is off center and, indeed, its location even slowly
moves! If an animal knew the direction of both geographic and
magnetic north, it would have a measure of its displacement
east or west; the angle between the two poles varies as one
moves east or west.

To date, there is no convincing evidence that animals use
the magnetic field for anything more than a simple compass.
But it is worth noting that the magnetic field of the earth does
contain "map" or position information if any animal could
make use of it. We do not know if they do, and thus the sensory
basis of animal navigation remains one of the more puzzling
questions in animal behavior.   cwt

Pigeon Homing
How do they navigate?

Much of the experimental work on bird navigation has been carried out with homing pigeons. It is possible to transport these birds long distances in any direction from the lofts where they live, and they promptly fly home again. Thus, the homing of a pigeon is quite analogous to that of the Laysan albatross or the Manx shearwater, but pigeons are much easier to work with. Thus, a great deal is known about pigeon homing.

Pigeons, like migratory birds, seem to use a map-and-compass strategy. They prefer to use the sun as a compass and are able to compensate for its movement across the sky. Under overcast skies, when the sun is not visible, pigeons switch and use the earth's magnetic field as a compass. Surrounding the pigeon's head with a pair of coils, it is possible to alter the magnetic field by passing a small current through the coils. By varying the direction in which the current flows, the polarity of the magnetic field can be changed. If a pigeon is released under overcast skies with the normal magnetic field, it simply flies home. But a pigeon with the reversed field frequently flies in exactly the opposite direction away from the home loft. Under sunny skies, the same coils have almost no effect.

The "map" which pigeons use to locate the loft is still not fully understood and various theories as to its nature have been proposed. One feature of their environment that the birds certainly use is familiar landmarks in the area surrounding their home. As they get close to the loft, pigeons will often turn suddenly and fly directly towards it when they get in the region that they know well. But pigeons can home from far beyond this area, so their familiarity with the terrain can only account for the last leg of their flight.

What could it be that they use at greater distances? A popular early idea was the sun might be involved much as it has been found to be in telling them compass direction. Determining position from the sun is much more tricky than using it as a compass, however: it is for this task that humans use a sextant. Experiments to see whether pigeons had some equivalent capacity have proved negative. Indeed, it has been found that birds wearing frosted contact lenses, through which the sun would only appear as a vague blur, could home to within a kilometer of their loft. Clearly, they could get that close to home without the aid of the sun or of landmarks.

Another possibility that has been suggested is that they use "route reversal," telling where they have been taken to by working out exactly where the lorry went on its outward journey. In theory it would be possible to work out where the destination was by analyzing all the twists and bends on the way, as well as the lengths of the straights between them, and adding up all of these. Some experiments suggest that pigeons do tend to leave their release site in the opposite direction from that in which the lorry carrying them left home. But they certainly do not retrace its whole route exactly and, because pigeons which were anesthetized for the outward journey can home perfectly well, this theory is not a popular one.

A more recent suggestion is that pigeons use familiar odors to find the loft, not as salmon do, but in a much more complex way. The proposal is that a pigeon in its loft learns the odors that are brought from each direction by the wind. Thus, when a pigeon is released, it recognizes the odor and remembers the direction of the wind that brought it to the loft. It then knows in which direction home lies. Additionally, the pigeon may sample the odors on the way to the release point and then use that sequence of odors as a set of guide posts to find the loft. For example, if pigeons are raised in a loft in which the direction of the wind is altered by large deflecting panels, there is a corresponding deflection in the pigeon's orientation. However, the pigeons show a clear deflection on sunny days, but orient towards home under overcast conditions so the smell hypothesis cannot be the whole story. Furthermore, pigeons which, due to an anesthetic, cannot smell anything still orient normally. Thus, the smell theory is a highly controversial one.

Even more contentious is the idea that pigeons might use

the earth's magnetic field to find their loft. The magnetic field could be used to obtain position information but, to do so, would require extraordinary sensitivity to magnetic fields. Yet there is evidence that pigeons could detect such small fields.

One suggestion comes from releasing pigeons at places where the earth's magnetic field is distorted by large deposits of magnetic iron. Pigeons released at such "magnetic anomalies" are disoriented until they reach magnetically normal terrain, where they head for home. The accuracy of the pigeon's orientation can be plotted against the distortion of the magnetic field and it seems that small distortions do disorient the birds. This is a hint that pigeons may be responding to remarkably weak magnetic fields, fields far weaker than would have any effect

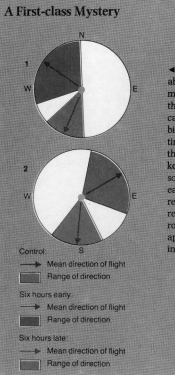

A First-class Mystery

◄ **Sun compass.** Pigeons are able to compensate for the sun's movements and use the sun to give them a compass to navigate by. This can be demonstrated by tricking the birds into thinking it is a different time of day and then releasing them. Those that were tricked were kept under artificial light conditions so that the day occurred (1) 6 hours earlier or (2) 6 hours later than it really was. When the birds were released north of their loft they flew roughly west or east compared to approximately south for those kept in normal daytime.

Control:
→ Mean direction of flight
▮ Range of direction

Six hours early:
→ Mean direction of flight
▮ Range of direction

Six hours late:
→ Mean direction of flight
▮ Range of direction

◄▲ **A flock of Racing pigeons**
takes to the air on the long journey
to their home lofts.

on a magnetic compass. Thus, there is a suggestion that
perhaps magnetic fields play some role in the "map."

Finally, it is worth mentioning that the sensory world of the
homing pigeon is quite different from ours. Where our hearing
stops at a frequency of about 20Hz, the pigeon can hear frequencies as low as 0.1Hz or 1 cycle every 10 seconds! Pigeons
can also see in the ultraviolet, can respond to the polarization
of light and are exquisitely sensitive to substrate vibrations. All
in all, the humble pigeon is quite an extraordinary animal, with
many sensory capacities.   cwt

▼ **Finding the way home.** Pigeons
are able to determine compass
directions using the sun and/or the
earth's magnetic field, but when
released how do they know which
direction is home? Do they use:
(1) the sun; (2) route reversal;

(3) odors; (4) the earth's magnetic
field; or (5) familiar landmarks near
home? All these theories have been
tested and some evidence favors
each of them. They certainly do not
use one method to the exclusion of
the others.

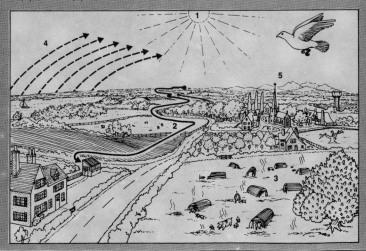

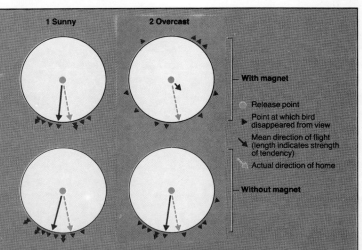

With magnet

● Release point

▶ Point at which bird
disappeared from view

➔ Mean direction of flight
(length indicates strength
of tendency)

Actual direction of home

Without magnet

▲ **Magnetic compass.** The fact that pigeons can also use the earth's magnetic
field to navigate by can be demonstrated by confusing the birds by placing
a bar magnet on their backs which produces a stronger and different field from
the earth's.

However, it is clear that on sunny days (1) there is no effect on their
direction with the magnet in place (ie the sun compass is being used) whereas
(2) on overcast days with the magnet in place the birds generally lose all sense
of direction (ie the magnetic compass is used—but confused).

Animal Relationships

Few animals live out their lives entirely in a solitary state. Apart from all else, most animals reproduce sexually and this requires a partner, except in the odd cases of hermaphrodites that fertilize themselves and some of the sessile marine animals which release their sperm and egg cells into the open ocean. Thus, reproduction is one reason why animals form relationships with one another, the subject of this section.

Here we move on to talk about the behavior patterns that animals show towards one another. Even the simplest of animals communicate with one another, if only by secreting chemicals that attract or repel their kind. Animals may communicate about many things, but two which deserve special attention are courtship and aggression. Displaying and fighting spread animals out so that they do not compete with one another; courtship brings them together so that they can mate. These are sweeping statements, however; perhaps, in reproduction more than in any other way, animal species differ enormously in the behavior that they show. The articles on parental behavior and on breeding systems give a flavor of that variety.

◀ **Aggressive encounter** between two Silverbacked jackals (*Canis mesomelas*).

COMMUNICATION

Warning coloration in insects. . . Distraction displays in birds. . . Facial expressions in monkeys. . . Types of signal and their meaning. . . Attracting mates and threatening intruders. . . Communication within groups. . . Courtship displays. . . Mediums of communication—through sound, visual signals and smell. . . Songs of whales, stridulation of grasshoppers. . . Color change in fish. . . Pheromones. . . Electrical communication in fish. . . Teaching chimps sign language. . . Fireflies. . . Scent marking. . . The dance language of honeybees. . .

THE singing of birds and the barking of dogs, the roaring of deer, the distinctive smell of a tomcat, the brightly colored patches on the wings of butterflies and on tropical fishes: all of these are examples of the signals that animals use when they communicate with one another. For prey animals to avoid being eaten, and for predators to creep up on their prey undetected, it is best to blend in with the background, matching its color and being as still and as silent as possible. So, when animals produce sights, sounds or smells that make them conspicuous, we can be certain that what they are making are signals important enough to them to be worth the risk. One of the great tasks of the study of animal behavior has been an attempt to understand why an animal produces the particular signals it does and what message they convey to the recipient.

In most cases, signals are used between members of the same species, but this is not always so. For example, the bright black-and-yellow stripes on a wasp are a warning signal, indicating to birds that might otherwise try to eat it that it is dangerous; and other brightly colored insects such as the Monarch butterfly, are so because they taste unpleasant. After one or two experiences, birds learn to recognize and to ignore them. Thus, both predator and prey avoid an unpleasant experience. But, although receivers may benefit by getting useful information, all that is necessary for communication to occur is that it is worthwhile for the animal producing the signal.

Animals also sometimes deceive one another when they communicate. Indeed, it has sometimes been argued that all communication consists of one animal manipulating others for its own ends. But this is putting it too strongly: in most communication, accurate and useful information passes from one animal to another. The main cases where this is is not so are, once again, instances where communication takes place between different species. Mimicry, as in the Viceroy butterfly, which looks almost identical to the distasteful Monarch even though it actually tastes nice, is one example. The distraction display of a mother plover, leading a fox away from her nest by pretending to have a broken wing, is another. Other animals appear bigger than they are and so frighten predators. The Eyed hawkmoth will suddenly reveal two large eye-spots on its hindwings which are so far apart that, if they were eyes, they could only belong to an animal at least large enough to eat the approaching predator. The Pearl-spotted owlet actually has markings that simulate a face on the back of its head. If you have ever looked into a nest box in which a Blue tit is sitting on her eggs, you too will have had a distressing experience.

▲ **Evil eye.** At rest this moth (*Automeris* species) is well camouflaged TOP. If disturbed by a predator, however, it moves its front wings forward to reveal bright markings on its hindwings BELOW that are normally kept hidden. These look like the eyes of a creature of considerably greater size and ferocity than a moth.

▶ **Roaring challenge.** In the rutting season, Red deer (*Cervus elephas*) stags roar to ward off rivals.

▼ **The false face** on the back of the Pearl-spotted owlet's (*Glaucidium perlatum*) head presumably deters attackers from the rear.

The bird spreads its wings wide and hunches its shoulders, hisses and rattles the wing feathers together. It is hard to believe it is not an altogether more formidable creature.

Most signals are concerned with a few rather general sorts of message and do not have very precise meanings as words in our language do. They are much more like the expressions we make with our faces: the smile of happiness, the frown of displeasure, the raised eyebrows of recognition. Like these human expressions, the signals of animals may be graded, a deep frown or a loud bark indicating greater displeasure than a lesser one. But this is not always so, and it is mainly the case in sophisticated animals, such as monkeys, which live in tight social groups. The signals of many other animals are very stereotyped and most cannot transmit these shades of meaning. There are good reasons for this. One is that different signals must be clear and distinct from one another to avoid being confused and to stand out against all the other sights, sounds or smells in the environment. A highly constant signal, perhaps repeated many times (introducing what is called "redundancy"), makes sure that the message gets across. A second reason for signals not varying much is that many of them carry a constant message such as species and sex. So a male stickleback on his territory always reveals a bright red belly and an iridescent blue face. It has also been suggested that fighting

animals should give the strongest possible signal because this makes it most likely that the rival will retreat in terror. Thus, the signals with which fighting animals threaten one another are rather strong and constant, giving little indication of how much they really want to win. Like wage bargainers, they thump the table and demand the maximum, though they would actually settle for much less!

One signal can mean different things to different listeners. Although it may be very constant, the song of a bird singing on his territory usually has a variety of information coded into it. It will certainly indicate which species he belongs to, as any bird watcher knows, and the very fact that he is singing is likely to mean that he is a male and is in breeding condition. His song may include personal idiosyncrasies: he may have a different song that he sings when he spots a rival; he may belong to a species in which males stop singing when they have a mate. A short snatch of a song indicates a great deal, and we can tell from their reaction whether other birds act on the information it contains. One way of doing this is to play tape recordings of songs to birds and see how they react. Territorial males usually get angry, but more so with songs of strangers than with those of their neighbors to which they are already accustomed. So, the fact that song varies from bird to bird enables them to differentiate between individuals. If songs are played on an empty territory, male birds tend to keep out of it, and females of some species actually show the mating posture when songs are played to them, even though they are entirely on their own. From experiments like these we can tell that the meaning of song differs between hearers: it attracts females and repels males.

The example of bird song indicates clearly two of the main functions of animal signals: the repulsion of rivals or competitors and the attraction of mates or other companions. Animals that have territories on which they feed or those, like stags, that maintain a harem by fighting off other males, must continually advertise their presence to keep intruders at bay. If an intruder does appear, an elaborate repertoire of signals may be used between the rivals, which enables them to assess their chances without the spilling of blood. In birds, such signals or displays are often visual and we can tell from what follows what they indicate. A display followed by attack would signify extreme aggression whereas one followed by retreat would be a submission signal. Deer have roaring matches in which rivals march up and down beside each other, making a tremendous noise: the winner is the one with the greatest staying power. The same animal would probably win if they did actually fight.

Many animals also have a selection of different courtship displays which are used to attract and excite females. Here it is often not easy to assign a particular message to each. Instead it seems that the male that can display in the most interesting,

▲ **Facial expressions in wolves**
(1) Friendly. (2) Submissive.
(3) Playful. (4) Very offensive.
(5) Very defensive.
(6) Aggressively defensive.

▶ **The luxuriance of courtship display.** The head of the Great argus pheasant (*Argusianus argus*) peers out from his gorgeous plumage, with its hypnotic whirlpool eye patterns. Its courtship ritual is as elaborate as its visual display, involving clearing a dance-floor of 4–5m (13–16ft) radius and soliciting the female with calls, foot tapping and leaping in full regalia.

▼ **Chins up.** An adult Chinstrap penguin (*Pygoscelis antarctica*) calls and displays to its chick on return to the breeding site.

exciting and varied way is the one that is most attractive. So different signals follow each other thick and fast in a riot of posturing.

Animals that live in social groups may transmit to one another quite subtle messages which help to pass information around the group and to coordinate their movements. Feeding chimpanzees, for example, have a particular call which attracts others to a source of food. Moving animals may make soft sounds, or have small visual signals, such as the black patch on the back of a lion's ear or the wing bars of many birds, which help to maintain contact between the members of the group. If a group is threatened, alarm signals are produced. For example, many birds have a thin, high-pitched "seep" call that they emit in danger. This call has characteristics which make it extremely hard to locate and birds can thus alert others

without making themselves vulnerable to predators. The call is similar in several species and reveals little about the sender. The receiver does not need to know the species of the sender; indeed, it would be much more relevant to know the species of the predator! While birds often do have several alarm signals, these are not specific to particular predators. However, Vervet monkeys do use different calls for a snake, a leopard and an eagle, the hearers reacting quite differently to each and in a way appropriate to the threat in question. This is about as close as animal communication comes to producing words: the calling animal is, in effect, shouting the name of the predator it has spotted.

Whatever the message that is encoded in a signal, most animal signals take the form of a sight, a sound or a smell. Each of these three "channels" of communication has different properties which suit it to particular uses. For example, sound waves radiate out rapidly in all directions and can be heard a long way off. They travel even more rapidly and for greater distances underwater: it is estimated that the song of the Humpback whale can be heard at least 1,200km (750mi) away. These properties make sounds ideal for use as advertising signals, such as those used in mate attraction. Sounds are also best for transmitting a great deal of complicated information quickly: it is doubtless for this reason that our language is based on noises rather than on gestures or on smells. We can bombard one another with rapidly changing messages in a way

that would be impossible using other means. Our vocal chords, and a mechanism in the windpipe of birds called the syrinx, provide particularly flexible means of producing highly varied sounds in quick succession. But there are many other ways in which animals make noises as signals: rabbits thump their feet on the ground, snipe produce a drumming sound by passing air rapidly through their tail feathers as they fly, grasshoppers "stridulate" by rubbing their hind legs against their wings, and many frogs have a large vocal sac which amplifies the sounds they make.

Visual signals travel even more rapidly than sound but, unless an animal is enormous, they cannot be seen far away, nor do they travel around obstacles. Thus, they are not very good for long-range communication. But they are excellent for rather private communication over short distances. Visual signals are ideal and often used for courtship displays: two animals showing off their color patterns to one another in the undergrowth have little danger of attracting a predator. They cannot say very complicated things to each other, at least not at speed, but they can produce simple and unambiguous messages by gestures or by color patterns. Some animals even change color to give a broader range of signals. Many male birds become more highly patterned in the breeding season, molting their feathers to achieve this; at around the time they ovulate, some female monkeys have a swollen and brightly colored rear as a signal to attract males. Most strikingly, many

Fireflies

Animals that use visual signals usually communicate during the day and rely upon the light from the sun to illuminate their displays. But fireflies are among those species than can communicate by night because they emit light of their own. (RIGHT). They are remarkably efficient light emitters, unlike most natural and man-made light sources, which also generate heat. The males fly around, producing a pattern of light flashes which is characteristic of the species to which they belong. The females have a different flash pattern with which they reply, and this enables the male to home in on them in the dark to mate. For example, in one species, *Photinus macdermotti*, the male makes two flashes about two seconds apart and, if the female detects them, she will reply with a single flash about one second later. The female only replies if the male's signal is exactly right. So, in a locality where there may be many species, these simple code-like signals can enable individuals to ensure that they obtain

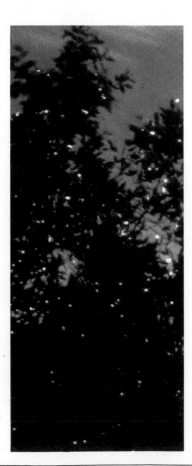

mates belonging to the same species as themselves. There are, however, some fascinating additional complexities in the situation. The females of a predatory firefly called *Photuris versicolor* sit in wait until they detect the flashes of the male of another species and then mimic the flash pattern of the appropriate female: for example, they give a single flash to the *Photinus* male described above. The unfortunate male flies in to mate and ends up being eaten. Not surprisingly, the males of the several different species that *Photuris* females can imitate have become rather wary in their approach to prospective mates! What of the *Photuris* male? He too can imitate the flashes of other species, but seems not to use this ability to find prey. Instead he waits for an answering female, flies in close to her and then switches to the pattern of his own species. In this way he discovers a female with whom he can mate and avoids the unpleasant end that she reserves for the males of other species.

▶ **Night-time croak**—a tree frog calling ABOVE. The importance of sending strong signals is brought home by the gross distortions often necessary to produce them. The huge vocal sac of this frog acts as a resonator, enabling it to be heard up to ten times farther away than a frog without such an apparatus. This species is *Hyla ebraccata*.

▶ **Wing-bars,** as displayed by this chaffinch (*Fringilla coelebs*) BELOW in flight, help to maintain contact between members of the same species. These bars are visible when the more obvious breast plumage is not.

◀ **Flashing lights** of fireflies displaying at night.

fishes can change their patterning over a very short period: the male Bull's-eye fish, for example, is silver when afraid, orange when aggressive.

The last of the three main senses, smell, might seem a very poor means of communication. One certainly could not transmit a lot of complicated ideas rapidly by this means for, once a smell has been broadcast, the animal has to wait for it to diffuse away before a different one can be released. Otherwise the different messages would all become jumbled together. But, by using specific chemicals known as "pheromones," animals can code quite complicated messages in smells. Not only can they indicate which species they belong to, but sometimes also which individual they are, as in mongooses, or how aggressive they are, as in some fishes. Moths use smells to advertise for mates, the female releasing a tiny quantity of pheromone which is wafted downwind from her and may be detected by males several kilometers away. How far it spreads, and in what direction, depend on the wind but, if conditions are good, it can be detected much farther away than could any sound or visual signal produced by such a tiny animal. The male is supremely sensitive to that exact chemical and flies upwind to home in on the female and start his courtship.

Some animals use channels of communication unknown to

Scent Marking

Smells can be used for several other purposes besides mate attraction. For example, male Ring-tailed lemurs involved in a "stink" fight rub (**1**) the secretions from a gland on their forearms on to their huge bushy tail and then shake the tail rapidly backwards and forwards to waft the smell to their rival. Skunks (**2**) have their notoriously unpleasant smell with which to drive off predators. In the honeybee, foragers mark sources of food with scent from a gland in their abdomen, and this helps others to find the site. Likewise, ants (**3**) lay odor trails which are faithfully followed by others on the way to and from sources of food. These last two examples illustrate a unique property of communication by smell: that it can be used even when the signaler is not present because the signal persists for some time after an animal has marked an object with its smell. Although insects, such as honeybees, have several different glands which they use to deposit various odors, the use of smell in communication is most sophisticated in mammals. For example, with our own comparatively poor ability to smell, it is hard to imagine just how detailed and fascinating the region at the foot of a lamp post is to a passing dog! Dogs and other canids, such as foxes (**4**), use urine to mark their territories, and the scent probably indicates individual identity as well as how long ago it was that the mark was made, because the chemicals involved slowly break down and disperse. Other animals, such as rhinoceroses (**5**), mark their territories with piles of feces, left in particular places, especially round the edges where other individuals may encroach. Special glands, sometimes around the anus, as in the Tasmanian devil (**6**), but also in other places such as on the chin, the legs or between the soles of the feet, as in hyenas (**7**) or near the eyes in deer, may also be involved in producing scents which are rubbed on to objects or other animals. Mammals, such as Bighorn sheep (**8**) often sniff and smell at each other and, when they do so, they can probably tell a great deal about the other individual, such as whether or not it is a relative or, if it has been marked by others, which group it belongs to. The world of smells is complex indeed, and we are only just beginning to understand it.

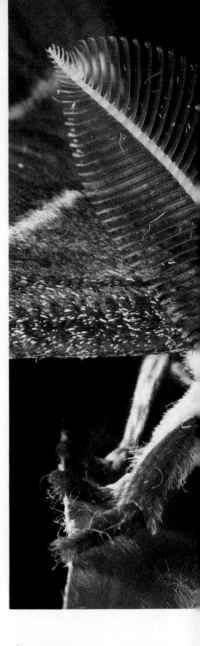

humans, the electrical communication of some fishes being the most striking example. Such fishes have nerve or muscle cells specialized to function as batteries. Some, such as the Electric eel and Torpedo ray, can produce sufficient power to kill prey, but electricity is used more often to learn about the environment and to communicate. The signals may be fluctuating continuous signals or intermittent pulses, and the frequency of peaks or pulses can convey not only the identity of the species but can attract females and indicate aggression, as with more conventional means of communication.

Although most animals have many different signals at their disposal, none has quite the complexity of human language. Monkeys and apes come closest to us with their facial expressions, and do produce various grunts and other noises, but certainly nothing like speech. The shape of their mouth, throat and vocal chords is quite different from ours so they are incapable of producing the rich variety of sounds that we can. A chimpanzee called Viki, brought up in a human family and treated just like a child, ultimately learnt to say "mama," "papa," "cup" and "up." Was this just because she could not manage the sounds necessary to speak more? This is almost certainly so. Another chimp, Washoe, was taught Ameslan, a sign language based on gestures which is used by the deaf in America. She mastered nearly 200 words and used them in the right way; she would also string them together in rudimentary sentences, for instance, saying to a visitor "You— Washoe—go—airplane." Being used to cans containing soup, she insisted to one of her companions that a soft-drink one had food inside it and a furious argument developed between them as a result. Other chimps have equally impressive achievements to their credit, although none has learnt as much or as quickly as a human could. However, what they can do suggests that their own natural communication systems, and those of other animals, may turn out to be more complicated than we assume.

PJBS

▲ **Pheromone detectors.** The antennae of this male Atlas moth (*Coscinocera hercules*) can detect minute quantities of a chemical released by the female when she is ready to mate.

▶ **Leaving a smell**—scent marking. This White-tailed deer (*Odocoileus virginianus*) buck is using a gland situated near the eye to mark out his territory, which will warn other bucks of his presence in the vicinity.

Buzzing Bees

The dance language of honeybees

Communication is essential even to most solitary species: it serves to bring individuals together for mating and, often, to enable parents to take care of their offspring. Among group-living animals, of course, far more communication is usually necessary to make the society work. The most highly social animals in the world are found among the termites and Hymenoptera (ants, wasps and bees), and their lives are clearly dominated by chemical, touch and sound communication.

By far the most outstanding example of insect communication is the dance language of honeybees. When a forager bee returns to the hive from a good source of food, she will perform a dance. The dance is usually in the form of a compressed figure-of-eight. During the central portion of this maneuver, the dancer runs in a straight line and vibrates her body rapidly from side to side. At the same time she vibrates her folded wings to produce a rumbling buzz. Other bees crowd around the dancer as if to listen. In 1944, the great German zoologist, Karl von Frisch, discovered that this straight portion of the dance—"the waggle run"—specifies the distance and direction of the food source.

Von Frisch already knew from his own work 30 years earlier that the dances stimulate other bees to search for food, and that the "recruits" learn the odor of the flowers by smelling the waxy fur of the dancer's body which readily absorbs scent. Moreover, dance attenders will "ask" for food samples by means of a particular sound, and so learn something about the quality of the food being advertised. Now he noticed that the waggle runs, which are usually performed on the vertical sheets of honeycomb in the hive, point upwards when the food is in the direction of the sun. He saw, too, that a food source located, say, 80° to the left of the sun's azimuth (horizontal direction) results in dances 80° to the left of vertical. Hence, vertical is taken as the direction of the sun, and the waggle runs are aimed left or right of vertical by just the angle between the food and the sun's azimuth outside.

Von Frisch also noticed that the number of waggles increased as the food was moved farther away. Each waggle corresponds to roughly 75m (250ft) near the hive, and declines in value slowly to about 50m (165ft) each as the food is moved beyond a few kilometers. Hence, the distance to a food source can be read directly from the number of waggles (or the duration of the buzzing; we have no idea which measure recruits may use). Food located closer to the hive than 75m (250ft) is indicated by a so-called "round dance" containing no waggles or sound at all.

Von Frisch referred to these performances as a "dance language" because, like language as we intuitively think of it, the dances appear to communicate abstract information about a place distant in both space and time from the "speaker," and use arbitrary, mutually understood conventions. By conventions we mean defining "up" as the direction of the sun—it could just as easily have been "down" or, for that matter, 13° left of vertical so long as both sender and receiver agree—and defining the value of each waggle. The apparent arbitrariness of the distance convention is compellingly illustrated by the observation that each of the two dozen or so races of honeybee (as well as the three other species which live in the tropics of the Old World) define waggles differently. These "dialects" run

▲ **Types of honeybee dance.** Foragers returning to the hive perform dances of two general forms. LEFT the "waggle dance," which involves a figure of eight path. During the straight part of the dance, the dancer vibrates its body and potential recruits crowd around. The direction of the "waggle run" indicates the direction of the food source (see diagram below). The "round dance" ABOVE RIGHT contains no waggles and is used when food is located closer to the hive than 75m (250ft).

▶ **Bee dancefloor.** The central bee, indicated by an arrow, is performing a dance.

▼ **What does a bee dance convey?** The direction of the straight part of the dance when the bee "waggles" is directly related to the direction of food. For example, if the food source is in line with the sun (**A**) the dance is upward on the comb or if (**B**) it is opposite the sun the dance is downward. (**C**) If the food source is 70° right of the sun, the dance is 70° right of vertical.

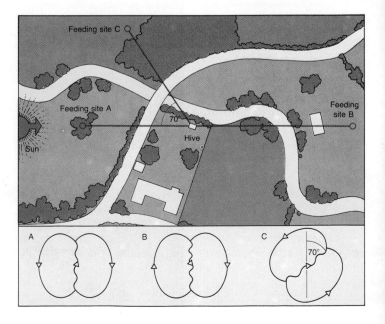

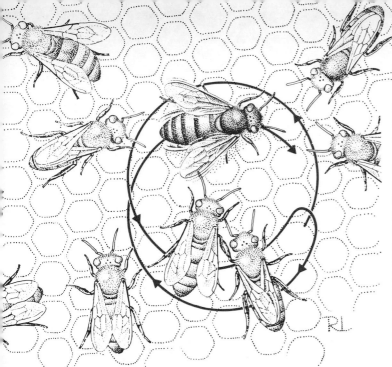

from 5m (16ft) per waggle to 75m (250ft). In terms of information content, the dance language would appear to be second only to human speech.

In the late 1960s, several researchers began to question whether von Frisch's work had really proven that the dance correlations actually communicate abstract information, and they performed experiments demonstrating that odor alone is sufficient to account for much of von Frisch's data. Foragers, they pointed out, not only bring back floral scents, but local odors which may be familiar to other bees. In addition, foragers have scent glands with which they can leave odor marks in the field. Subsequent experiments, however, in which visiting foragers were tricked into "lying" about either the direction or the distance of the food source they were visiting have proved that the dance works: most recruits went to the location specified by the waggle run rather than the spot actually being visited by the dancing forager.

The dance language provides honeybees with a tremendous advantage over most other nectar-gathering insects. A single, successful scout bee can, if the discovery is sufficiently outstanding, recruit several foragers that, upon their return, will dance and recruit still others, and so on. It is not unusual to see the number of bees at a food source double every few minutes such that, if the patch of flowers (or, in experiments, the dish of sugar water) is not saturated, 1,000 of a typical colony's 10,000 foragers could be recruited within an hour. This sharing of information amplifies greatly the food-gathering efficiency of the colony.

Subsequent work by von Frisch's students and others has shown that the dance can also be used to muster bees for other purposes. Sources of pollen (which provides bees with protein), water (which is used to cool the hive in hot weather) and propolis (tree sap, used as mortar to narrow the entrance to the hive, fill cracks and encase objects in the colony too large to be removed) are all indicated by dancing. Perhaps the most remarkable use of the dance, however, comes during swarming. Honeybee colonies reproduce by "fission:" the queen and about half of the workers leave in the late spring to find a new home. (A new queen has been reared in the old hive to replace her.) The departing bees form a swarm on a nearby tree limb while scouts search for suitable nest cavities. Discoveries are reported via the dance language, and scouts regularly compare the sites advertised by other bees with their own discoveries and subsequently dance for the most suitable alternative.

Careful studies have shown that scouts evaluate the size of the cavity—10 litres (18 pints) is about right—its distance from the home colony (it should be neither too close nor too far away), its height above the ground—3m (10ft) is considered good—the direction of the entrance, the cavity's freedom from drafts and leaks and so on. After two or more days of "debate" among the scouts, a concensus is reached and the swarm departs for its new home. Obviously, the colony benefits greatly from this ability to exchange information about what is a very rare and essential commodity: suitably snug nest sites. That bees decide "democratically" seems as incredible today as when it was first discovered, and serves to remind us of the ability of natural selection to program highly complex behavior into even very small brains.

JLG

AGGRESSION

Defending food and breeding rights from competitors...
Protecting territories from rivals... The Red deer rut...
Hartebeest leks... Elephant seal harems... Fighting to the
death... Bison and baboon herds... Elephants in musth...
Female conflict in monkeys... Alliances in turkeys and
lions... Animal weapons—horns and teeth... Fighting by
convention—bluff and counterbluff... Lorenz on
aggression... Friends and enemies...

THE idea of "Nature red in tooth and claw" permeates our folklore. Not surprisingly, perhaps, for many species possess teeth and claws of quite formidable dimensions capable of dealing fatal injuries to an opponent. Everyone knows that dogs will attack, even kill, each other when they meet for the first time. And many a tomcat has appeared with bloody nose and torn ears the next morning after a night out on the tiles. But how often do animals fight to the death? What is all the fighting about, anyway? And why is it always tomcats that get the bloody noses, seldom the females.

Most fighting occurs over access to the resources that an animal needs to be able to live, breed and rear its young. Such resources can include individual items of food or whole areas with the food they contain (that is, territories), waterholes, safe sleeping or nesting sites, mating or egg-laying sites or even individual mates. So long as there is a surfeit of any given resource, competition for it will be negligible. If a resource is scarce, however, pressure of competition will lead to contests for its

Fighting by Convention

Contests in which the opponents actually fight each other are likely to result in injury to one or both parties. There is little point in risking injury unnecessarily, so that many species have evolved ways of settling disputes by means of a convention. By performing some risk-free activity, an animal may be able to convey information about its prowess and thus persuade its opponent to abandon the contest rather than risk injury in a fight it is likely to lose. Male monkeys and hippos yawn to reveal the dangerous canines that they use in fighting. Such displays need not reflect fighting ability directly, though there will usually be some correlation with stamina or body size, both of which influence fighting ability.

Siamese fighting fish swim parallel to each other, beating their fins in unison in a kind of war of attrition. Deer stags roar at each other, increasing the frequency in step as the bout progresses: larger animals can generally make louder noises at lower pitches than small ones, so roaring may convey information about body size.

Could animals use these conventions to bluff each other by pretending to have more fighting ability than they really do? Almost all do to some extent. Cats and dogs make their fur stand on end, so appearing to be larger. Even humans feel a prickling in the nape of the neck in similar situations—a leftover from when our ape-like ancestors had fur that could be erected. Such tactics, however, are only useful up to a point. Ultimately, fighting ability depends on size, and a small animal pretending to be very large would be unconvincing to a large one which could pretend to be larger still. Moreover, if too many individuals cheat too often, any animal that called its opponent's bluff would win disproportionately often, thus counterbalancing the advantages of a little cheating.

possession—if only because exercising physical power is one of the most effective ways of keeping the opposition at bay. The more important a resource is to an animal's ability to reproduce successfully (and dead men neither tell tales nor reproduce), the more fiercely it will fight for it.

The consequences of this are seen in the way a robin or a Blue tit defends its territory. To the Blue tit male, his territory represents the food that he and his mate will need to rear their voracious young successfully. If the territory is too small or has too poor a food supply, the nestlings will starve to death. Territories of this kind that are vigorously defended against other members of the species are common among species that form stable monogamous pairs, including gibbons, the tiny South American marmoset monkeys and many small antelope.

Territories need not always provide food. To the stickleback male, the territory that he defends so assiduously on the stream bed is like a stage from which he attracts females to lay their eggs in his nest rather than those of his rivals next door. The leks of the Sage grouse and the ruff consist of small territories, often only a few square meters in size, packed together in display areas like the booths in a market. They function solely as mating sites, since the females, once mated, fly off to lay their eggs in their nests elsewhere and care for the young alone. A male that does not own a territory on the lek will not usually be able to breed, and he can only acquire one by fighting other males for it.

The cost of fighting to gain and then keep control of a territory can be considerable, especially if the number of males wanting territories is very large relative to the number of territories available. Hartebeest also have a lek system, and the males fight their way through the less desirable peripheral territories to gain control over the central territories where the females tend to congregate. They will lose the territory to another male if they leave it, yet the territory is too small to sustain them, so that the males progressively lose weight until finally forced to leave or risk starving to death. Under similar circumstances, Red deer males that lose too much weight as a result of spending too much time fighting during the fall rut may fail to survive the hardships of the following winter.

Elephant seals come ashore for only about one month each year. During this time the females give birth and then conceive next year's pup. The males establish territories on the beach, each containing the resting places of several females over which the territory owner has exclusive mating rights. The bulls engage in bloody slogging matches in which the biggest and most powerful gain the largest territories with the most females. As the density of males builds up, however, the competition among them can become so intense that the territorial bulls are unable to keep every intruder out at the same time. Instead, a bull will resort to defending an individual female that is on heat to prevent some young upstart from mating with her while he is busy chasing off other intruders.

▲ **Cheetah challenges cheetah**—a group of male cheetahs (*Acinonyx jubatus*) attack another nomadic male that has wandered into their territory on the Serengeti Plains, Tanzania. The value of these territories and of cooperation between individuals is underlined by the result—the intruding male (on the right) was finally killed.

▶ **Signalling different intensities** of fear or threat. Animals escalate an encounter by using threats of increasing intensity of aggression, or de-escalate using submissive signals of increasing fear in the same way, as shown here for cats.

Increasing threat

Increasing fear

Territorial defense need not always involve direct conflict. Pairs of gibbons and male colobus monkeys, for example, advertise their territorial status with special loud calls that reverberate around the forest, while at the same time leaping around the treetops in spectacular acrobatic displays. Males in neighboring groups are invariably stimulated to reply on hearing another displaying, and the whoops and roars can be heard spreading like waves through the forest canopy. Some species mark their territories by depositing scent from special glands on twigs and rocks round the boundary, or by piles of dung deposited at special sites. The message they convey is "Beware all ye who enter here!"

Not all species defend territories, of course. Many like baboons, wildebeest, bison and starlings, live in groups or herds that move in areas too large to defend. Such groups usually contain several adult males. Even though they do not fight to defend resources on a general level, they still compete among themselves, sometimes over individual items of food, at other times over receptive individual females. The amount of aggression seen again depends on the level of competition for the resource in question. In baboons, for example, the number of fights between males increases dramatically when a female comes into heat, but tends to decline again in proportion to the number of females on heat at the time because proportionately more males are able to monopolize a female.

Fighting often occurs in "muted" form that minimizes the damage to the combatants. Usually it begins with threats of increasing intensity, escalating to more damaging forms of contact aggression only if neither contestant is prepared to back down. In Gelada baboons, for example, encounters usually begin with vocal threats and stares given at a distance. These may be followed by an approach to closer quarters to stand, with hunched shoulders, giving higher intensity visual and vocal threats at the opponent, escalating to slaps at the ground and lunges at the opponent. Finally, if the opponent still refuses to submit, the two protagonists may resort to actual jaw-fencing, trying to slash and bite each other with their formidable canine teeth.

Many species that possess potentially lethal weapons have evolved ways of minimizing the risk of injury when they do come to blows. Species such as baboons, seals and hippopotamuses, that stab and slash with canine teeth, generally have shoulders and necks that are protected by long fur capes or thickened skin that take the worst of the blows and prevent the more vulnerable internal organs sustaining serious injury.

In fact, animals rarely fight to the death. Instead, one

ANIMAL WEAPONS

▲ ▶ **Fighting** has often led to the evolution of weapons that improve fighting ability and give the possessor a better chance of winning. The shape of most weapons is intimately related to the species' style of fighting. When ibex (*Capra ibex*) butt each other (**1**), the blows are caught horn on horn, thus preventing damage to the skull and brain; their massive horns are heavily ridged to provide grip so that they will not slip off each other under impact.

Water deer (*Hydropotes inermis*) (**2**) use their canine-like teeth for slashing. The short sharp horns of a klipspringer (*Oreotragus oreotragus*) (**3**) are used for jabbing.

When deer fight, they try to push each other backwards off balance: the tines of the many branched antlers of White-tailed deer (*Odocoileus virginianus*) (**4**) interlock to provide a firm surface for pushing with.

The delicate reedbuck (*Redunca arundinum*) (**5**) engages in more genteel wrestling matches in which the contestants lock horns and try to twist each other's heads down on to the ground: their horns are short with a forward-curving tip that helps to prevent them sliding past each other.

Elephants (*Loxodonta africana*) have transformed their teeth (tusks) into battering rams (**6**).

Weapons are often costly to produce in terms of the energy needed to grow them. Heavy horns may also unbalance an animal on steep terrain or slow it down, making it more susceptible to fleet-footed predators. Thus, natural selection works against the advantages of growing ever-larger weapons. This results in an optimum weapon size for each species that reflects the way the weapon is used and the relative costs and benefits of weapons of different size in the particular social and ecological circumstances under which the species lives.

protagonist will usually turn tail and flee once it has decided that it cannot win, although, in some species, it can take several days of intense and often bloody fighting for a contest to reach this stage. One way of persuading an opponent to retreat without fighting is to appear to be superaggressive, especially in the case of very large species where animals are likely to cause considerable damage to each other if they do fight. Bull elephants undergo a state called "musth" in which they go mad and will attack almost anything on sight. Bulls in musth advertise their condition by conspicuous smell and have special glands on the temples that ooze profusely. Bulls in musth are extremely dangerous, and other bulls steer clear of them rather than risk a fight.

In the more social species it is essential that animals evolve mechanisms for defusing aggression, otherwise the social groups will disintegrate. In these species, disputes need to be carefully regulated to prevent them from getting out of control. Many species signal their submission in special ways. Dogs and other canids expose the neck as if to invite the victor to despatch them in the best traditions of medieval chivalry. Monkeys will cringe with characteristic screams, or present their rear to their opponent; the victor may touch, or even groom, the loser so as to re-establish cordial relationships and reduce tension.

Because much of the fighting among males occurs over access to sexually receptive females, the frequency of fighting often shows marked seasonal cycles, and is most common during the breeding season. Females of most species are in season only for a period of about one month each year. Just before this happens, the males, which have lived relatively amicably together until then, start to engage in a series of contests, testing their relative strengths against one another. Through this they sort themselves into a dominance hierarchy, with the highest-ranking males having priority of access to the best territories or to the individual females as they come into heat.

Many species of animals show this seasonality of aggression. It is controlled by the annual cycle in the levels of the male sex hormone, testosterone. Indeed, it has been found that animals that achieve high rank in their groups generally have higher levels of testosterone than those at the bottom of the hierarchy. This partly explains why males are usually more aggressive than females: females tend to have much lower levels of testosterone in their blood than males (though it may surprise most people to know that they have any at all!).

Nonetheless, females are not passive, unaggressive creatures. Those of most species are also organized into dominance hierarchies based on aggression, just as their males are. Although females may compete directly with males for food and other resources, in reproductive terms they compete most seriously among themselves. Aggression by females is usually much less spectacular than that among males and is therefore commonly overlooked. The consequences, however, may be no less serious. Just as weak males may end up at the bottom of the hierarchy and be unable to breed, so low-ranking females

▶ **Fatigue or threat?** A hippo yawn is not a demonstration of fatigue but one of intimidation, as he displays his arsenal of teeth to an opponent.

▲ **Slogging it out,** two massive male Southern elephant seals (*Mirounga leonina*) fight for the right to occupy part of a beach and its attendant females.

◄ **Turning white in fear**— dominance comes in all colors. Here a dominant (green) chameleon (*Chamaeleo senegalensis*) threatens a subordinate individual which demonstrates its lower rank by turning white.

► **Feathered fight**—two Great tits (*Parus major*) dispute over a food source.

Lorenz on Aggression

As a leading early ethologist, Konrad Lorenz's views on aggression were particularly influential. From his detailed comparative studies, he concluded that aggression served three main functions: the spacing of individuals in the habitat (especially in territorial species); selection of the strongest individuals for breeding through "rival fights;" and brood defense. This argument is based, however, on the idea that animals should act for the good of their species as is the suggestion that it is advantageous for only the strongest to breed. Now it is thought that natural selection acts mainly at the individual level.

Among Lorenz's many important observations, was the fact that the bright colors of many coral reef fish functioned as aggressive signals that kept members of the same species well spaced out without the need to expend a great deal of time and effort in fighting. Lorenz considered aggression to be the most primitive and powerful of the instincts, liable to erupt spontaneously in inappropriate ways if it was not given vent to naturally from time to time. To counteract these dangerously disruptive tendencies, he argued that the evolution of social living had necessitated mechanisms that inhibited aggression to prevent members of the same group killing one another. These

included submissive behavior and various forms of "friendly" or associative behavior, such as grooming in monkeys and the "triumph ceremonies" of his beloved geese. The more dangerous the species, he suggested, the more highly developed its bonding mechanisms. In humans, he saw this as leading to social conformity and dislike of outsiders. Being based on evolutionary and physiological mechanisms that are either false or oversimple, Lorenz's functional and motivational explanations are not now widely accepted. Nonetheless, many of his observations on behavior patterns and their origins have stood the test of time.

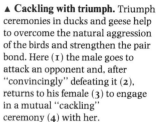

▲ **Cackling with triumph.** Triumph ceremonies in ducks and geese help to overcome the natural aggression of the birds and strengthen the pair bond. Here (**1**) the male goes to attack an opponent and, after "convincingly" defeating it (**2**), returns to his female (**3**) to engage in a mutual "cackling" ceremony (**4**) with her.

may find their reproductive outputs fall well below those of higher-ranking sisters. Among Talapoin monkeys and Gelada baboons, harassment by more dominant females generates stress in low-ranking individuals which disrupts their reproductive physiology; as a result, they fail to ovulate and, consequently, have lower birth rates than higher-ranking females.

In an attempt to overcome the disadvantages of low rank, female monkeys often form alliances, typically between close relatives. In this way, mothers, for example, are able to support their maturing daughters, with the result that an immature female may outrank older, more aggressive females because of the threat of intervention by her mother. Males may also form coalitions, sometimes of unrelated individuals; by acting in support of one another, the members may be able to acquire higher ranks than each would have been able to achieve alone. Among Tasmanian native hens and wild turkeys, brothers form alliances and hold joint territories, while lions are able to hold prides of females for longer if they form coalitions than if they try to do so alone.

Although in most species the male is bigger and more aggressive than the female, this is not necessarily always so. Female hyenas and Squirrel monkeys, for example, are slightly larger and more aggressive than the males, which they are normally able to dominate. In certain species of birds, such as jacanas and phalaropes, where there is complete reversal of the usual sex roles, the males look after the eggs while the larger, more brightly colored and more aggressive females fight among themselves for control over the males' nests in much the same way as male seals do over females. **RIMD**

COURTSHIP

*Roles of males and females. . . Why is courtship necessary?. . .
Traditional breeding sites. . . Making yourself conspicuous—
Ghost crab burrows. . . Displays. . . Antagonism between
males—spider courtship. . . External fertilization in
sticklebacks. . . Frog calls. . . Whether to mate, fight or
flee. . . Displacement behavior in ducks. . . Mating seasons. . .
Fighting for the right to mate. . .*

It is early spring in the northern hemisphere, the woodlands
are ringing to the songs of birds and, from a nearby pond,
the rasping croaks of frogs can be heard. If lucky, you might
happen to see a male hedgehog running in circles around his
intended mate or a dog fox playing with a vixen. In the evening
clouds of dancing male mosquitoes and midges emerge and the
tiny light of a female glow-worm (actually a beetle) can be seen
at the bottom of a hedge bank. On the open fields the mad
March hares are preparing for the breeding season by frantic
chasing, leaping, or even "boxing" with their forepaws as male
and female rear up on their hind legs facing each other. But
what is all this furious activity for?

The term "courtship" refers to the behavioral interactions
that occur between males and females before, during and just
after the act of mating. In some animals courtship is brief and
perfunctory but, in others, it lasts for a long time and involves
vigorous and elaborate displays. Typically, the male is the more
active partner during courtship. It is usually he that initiates

▶ **The upside down display** of a
male Blue bird of paradise
(*Paradisaea rudolphi*) is one of the
most spectacular in the animal
kingdom. These birds are a fine
example of the power of sexual
selection through evolution to make
males as attractive as possible to
females (see box below).

Sexual Selection

Within a breeding population,
there is often intense
competition between members
of one sex for opportunities to
mate. Typically, it is males that
compete with each other for
access to females because,
whereas females typically need
only mate once to ensure that
all their eggs are fertilized,
males can enhance their
reproductive success by mating
with many females. Sexual
selection is the evolutionary
force that arises when some
males are more successful than
others in obtaining matings.
Those males that mate with
many females will tend to pass
on whatever characteristics
enabled them to do so to their
numerous offspring; those that
fail to mate will not pass on
their characteristics. As a
result, characteristics that
enable males to acquire mates
will rapidly spread through a
population, and will tend to
become more and more
pronounced over successive
generations.

Male competition may take

one of two forms. Males may
fight with each other over
females, in which case sexual
selection favors those that are
largest, strongest or which have
the most effective weapons.
This form of sexual selection is
thought to be responsible for
the antlers of deer stags and for
the fact that, in many
mammals, males are very much
larger than females.
Alternatively, males may
compete in a less direct way,
mating success depending on
how effective a male is in
attracting and stimulating
females. Thus, variation in
male mating success apparently
depends on preferences among
females for certain males rather
than others. This form of sexual
selection favors elaborate male
displays and those physical
characteristics that are
associated with them, such as
bright colors and elaborate
plumage. The courtship display
and huge tail of the peacock are
thought to have evolved
through sexual selection based
on female preference.

The theory of sexual selection
has been controversial since it
was first formulated by Charles
Darwin. While there has been
no disagreement over the idea
that male strength and
weapons have evolved because
of the advantages they confer in
direct competition for females,
many people have doubted that
female choice can have been
responsible for elaborate male
displays and plumage. There is
very little clear evidence that
females do prefer those males
that have the most exaggerated
courtship characters. A recent
study has, however, obtained
such evidence. Males of the East
African widowbird have
enormously long black tails
which make a displaying male
visible from over a kilometer
(0.6mi) away. The tails of some
males were cut short and the
removed sections were stuck on
to the ends of the tails of other
males. Males with artificially
extended tails attracted more
females than those with normal
length tails or those whose tails
had been cut short.

◄▼► **The varied patterns of courtship.** (1) Male and female mandrill (*Papio sphinx*)—the male is bigger and more flamboyantly colored than the female. (2) Drake mallard (*Anas platyrhynchos*) showing ritualized preening, with a female behind. (3) A pair of Black-headed gulls (*Larus ridibundus*) "facing away" so as not to show their dark faces which are normally associated with aggression. (4) Female (top) and male Three-spined sticklebacks (*Gasterosteus aculeatus*). (5) Courting Two-lined salamanders (*Eurycea bislineata*), with the male lacerating the female's back. (6) Male Sage grouse (*Centrocercus urophasianus*) displaying at communal mating site with females behind. (7) Male European tree frog (*Hyla arborea*) calling.

species, females appear to be much more cautious than males before they mate.

Each of the many and varied activities forming a part of courtship behavior has a specific function. Some behavior patterns serve in the location or attraction of a potential mate, often over a considerable distance. Once a male and female have come together, the male commonly displays to her and his behavior serves to make her sexually receptive. In many species, the mating act is a complex affair and requires both partners to behave in very specific ways at particular moments if it is to be successful; the function of many courtship activities is to achieve precise synchronization of male and female activities. Finally, mating will only be successful in the long term if it occurs between males and females of the same species. Hybrid offspring are typically inviable or infertile and so represent a loss of reproductive potential for their parents. Many courtship activities ensure that individuals mate only with members of their own species. Thus, courtship fulfils four major functions: orientation; persuasion; synchronization; and reproductive isolation.

For small animals, especially those that normally live widely dispersed, finding and orienting towards a mate may pose enormous difficulties. In some species, the problem is resolved by males and females aggregating at some particular resource or at a traditional breeding site. For example, male and female dungflies are attracted to fresh cowpats, where mating occurs and in which females lay their eggs. Certain birds, such as the ruff and the Sage grouse, mate at communal mating sites called leks, which occupy the same site year after year. Likewise, toads tend to return to the same breeding pond every year. In many species, however, one sex produces signals that attract mates, sometimes over a great distance. Among many frogs and insects, the males make sounds that can be heard over hundreds of meters. The advertisement calls of frogs are produced by the vocal cords and are amplified by inflatable sacs in the mouth. Insects typically "call" by rapidly rubbing together parts of their hard external surfaces; grasshoppers rub their legs together, crickets their wings. The songs of some birds also work as auditory cues by which females locate males.

The fireflies are a group of nocturnal beetles that have special light-producing organs. Males fly around producing distinctive sequences of flashes, and females, sitting on the ground, respond to males of their own species by flashing in reply (see p66). Other insects attract mates by producing odors or pheromones. In many moths, females secrete odors at night; these are carried by the wind, and males, which have very acute senses of smell in their antennae, fly upwind to find them.

Certain animals modify their environment to make themselves conspicuous to potential mates. Male Ghost crabs dig a deep burrow and build the excavated sand into a pyramid that enables females to find them. Mole crickets call from the safety of a burrow which is specially shaped so that the sound is greatly amplified.

Among some animals, a meeting of male and female leads almost immediately to mating. In many species, however, one partner, usually the male, is more ready to mate than the other. He may then perform behavior patterns whose function is to stimulate the female until she is sexually receptive. Often, such

a sexual interaction and displays to the female whose behavior may involve little more than approaching the male if she is receptive, or moving away if she is not. This difference in male and female roles can be related to the fundamental difference between the sexes. Females produce large eggs in very small numbers compared to the millions of tiny sperms produced by males. As a result, each female is able to produce only a limited number of offspring whereas each male, if he could mate with many females, has the potential to produce an almost unlimited number of progeny. Thus, it is not surprising that, in many displays involve a variety of activities that stimulate the female in different ways. Male newts perform movements that produce a combination of visual, smell and touch stimuli. Many birds perform displays that combine visually striking postures with songs and calls. In Ring doves, it has been shown that exposure to male courtship induces the hormonal changes in females that make them sexually receptive and also leads them to ovulate. Some courtship rituals are extremely bizarre: males of the Two-lined salamander apply a secretion from glands on their head on to the female's skin. They then lacerate her skin with two protruding teeth so that the secretion, which can be regarded as an "aphrodisiac," enters her bloodstream.

For the male of some species, the female may be a severe threat, should she not respond to him sexually. She may attack him and, in some cases, eat him. In these circumstances, male courtship behavior may serve, not simply to stimulate the female sexually, but also to suppress non-sexual behavior. Many male spiders are markedly smaller than females and are at risk of being mistaken for a meal. In some species, the male vibrates the female's web with a characteristic rhythm, indicating that he is a potential mate and not prey. In other spiders, and in some insects, the male presents the female with a "nuptial gift," consisting of an insect wrapped in silk. While the female unwraps and eats the insect, the male is able to mate with her without being attacked. Among birds, males and females commonly show ambivalent responses to one another during pair formation, with tendencies to attack and to flee being apparent in their sexual behavior. During the courtship of Black-headed gulls, partners frequently "face away," concealing from one another their dark face patches which are prominently displayed in aggressive disputes.

Precise synchronization of male and female courtship activities is especially important in species in which there is external fertilization. Because they must meet outside the body, it is vital that eggs and sperms should be released at the same instant, lest they become dispersed before fertilization can occur. In the Three-spined stickleback, the female does not release her eggs unless the male nudges the base of her tail with his snout just after she has entered his nest. She then leaves the nest and he immediately follows her through it, shedding his sperm on to her eggs. Behavioral synchrony is also important in most mammals where the female must adopt a particular posture, called "lordosis" in rodents and cats, before the male can mount and mate with her successfully.

The role of courtship in ensuring that animals mate only with a member of their own species is typically achieved because courtship displays are highly species specific. Thus, animals produce odors, make sounds or perform visual displays that are distinct from those of other species. Allied to this, the partner that receives such signals is usually responsive only to the displays of its own species. In most frogs, males produce calls which, in terms of their pitch or timing, are very stereotyped and species specific. Females approach only the calls of their own species. In many frogs, the ears of females contain many sensory cells that can detect sound frequencies present in the male calls of their own species and very few that can detect other frequencies. Thus, female frogs may be more or less "deaf" to the male calls of species other than their own.

Mating Seasons

Most animals breed only at certain times of year, especially at high latitudes, where there are marked differences in environmental conditions at different seasons. The timing of breeding activity has clearly been influenced by natural selection, which has favored those animals that breed at such a time that their progeny are most likely to survive. In northern, temperate habitats, breeding typically starts in early spring so that the young gain the benefit of a rich food supply in spring and summer. In areas where climatic changes are irregular, breeding often does not occur at a specific time of year. In deserts, for example, breeding of many species follows heavy rain, whenever it happens to fall. In some large mammals which have long gestation periods, such as Red deer and sheep, the mating season occurs in the fall. The embryo develops over the winter and is born in the spring.

The fact that all members of a breeding population come into breeding condition at very much the same time of year is largely due to each individual responding in a similar way to changes in the environment.

The most important

environmental factors in higher latitudes are day length and, to a lesser extent, temperature. In animals that breed in spring, increasing day length induces the hormonal changes that bring about reproductive development. By contrast, for animals that have a mating season in the fall, such as deer and sheep, hormonal changes are stimulated by decreasing day length.

The onset of breeding is not, however, simply controlled by such factors as day length and temperature; for many animals these are augmented by the behavior of prospective mates. For example, female canaries come into reproductive condition more quickly if they hear male song than if they are kept in acoustic isolation from males. Female Green lizards do not develop fully mature eggs unless they are displayed to by males.

Breeding at the same time as others is advantageous as it leads to predators being "swamped;" there are so many eggs and young around that they can only eat a tiny proportion. By contrast, individuals that breed before or after the peak are much more likely to lose their progeny.

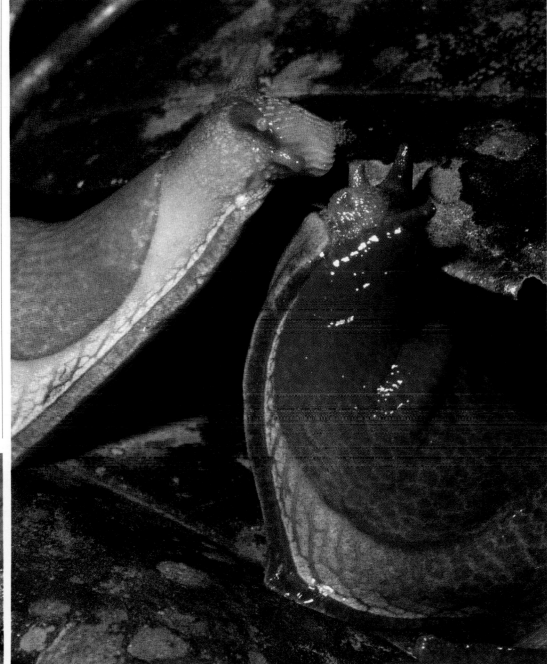

◄ **Necking—courtship in giraffes.** A bull giraffe (*Giraffa camelopardalis*) rubs the neck of a cow as part of courtship. During the breeding season a succession of bulls of increasing dominance will court a receptive cow so that by the time she is ready to mate the dominant bull in the locality is attending her and will be her mate.

► **Jungle courtship ritual.** Slugs and snails are hermaphrodite, but cross fertilization via mutual exchange of a sperm package is essential. Some slugs and snails have a so called "love dart," which is driven into the body of the partner and serves to stimulate copulation. Larger slugs, such as these tropical *Trichotoxon* species, have an elaborate courtship ritual which may last several hours before mating.

▼ **Smelly courtship ground**—a pair of Yellow dungflies (*Scathophaga stercoraria*) on a dungpat. Male and female are attracted to fresh dungpats where mating occurs.

The antennae of many male moths are selectively responsive to the odors emitted by females of their own species.

Of the many bizarre and strange aspects of courtship behavior, one that has attracted attention for many years is the ambivalent response that prospective mates often show to each other. As well as showing sexual responses, their behavior may contain elements of fear, aggression and other activities. It appears that courting animals, at least in the early stages of a sexual encounter, are in a state of motivational conflict, with tendencies to mate, to fight and to flee competing with one another. Many courtship interactions end quickly, before mating has occurred, apparently because some competing tendency is more powerful than the motivation to mate in at least one of the partners. When they are in this state of conflict, animals commonly behave in strange, apparently irrelevant ways. In many ducks, for example, a male will suddenly start to preen. Such behavior patterns are called "displacement activities." In many cases it appears that, during evolution, certain displacement activities have become more conspicuous and more stereotyped so that they now function as displays. These evolutionary changes are referred to as ritualization. During the courtship of Mallard ducks, for example, the male performs a display in which he turns his head and touches a conspicuous patch of feathers on his wing, called the "speculum," with his bill. This appears to be a ritualized form of displacement preening.

TRH

PARENTAL BEHAVIOR

Strategies with nil or minimal care... Mouth brooders in fish and frogs... Feeding nestlings and mammalian young... Number of young produced in swifts... How long to care for young... What makes young attractive to parents... Promotion of independence in monkeys... Parental experience in raising young...

Human parents and those of other mammals spend a lot of time and effort looking after their young. And yet many animals hardly care for their young at all. What they fail to provide in individual attention, they make up for in the numbers of young they produce, and each set of parents produces thousands, or even millions of fertilized eggs, very few of which survive to adulthood. Sea urchins, for example, waft clouds of eggs and sperms into the sea, but only a tiny number of the fertilized eggs that result will survive to breed themselves.

Some animals package the young in their eggs with a supply of food in the form of yolk; this helps them through the early stages of development. When young fish fry hatch from the egg, they still carry a yolk sac around with them, and can live on this for days. The yolk can be very large in reptiles, such as crocodiles and turtles, and, of course, in birds. A single ostrich egg contains about 800g (28oz) of yolk.

Some animals go further in their parental care, and ensure that the young continue to receive a good supply of food after the reserves in the yolk are exhausted. The female digger wasp *Sphex ichneumonias*, for example, digs a burrow and catches some Long-horned grasshoppers. She does not kill the grasshoppers but paralyzes them by stinging, puts them in the burrow, lays an egg on top of them and seals up the burrow. When her larvae hatch from the eggs, they eat this store of fresh food.

Mammals go to even greater lengths, carrying the young in their bodies until they have reached a stage when they ought to be able to fend for themselves. Among the placental mammals, mice carry the young in their womb for three weeks while, in elephants and humans, the young may continue to develop in the mother for many months. It is not only the mammals, however, that carry their young within their bodies. In some cichlid fishes, such as *Haplochromis* from the East African lakes, the female carries the fertilized eggs, and later the young fishes, in her mouth. The fry swim around the mother's head but they dart back into her mouth if danger threatens. Many frogs and toads keep the developing young in special cavities within the body.

After birth or hatching, the parents of many animal species continue to care for their young for days, months or even years, providing them with food, keeping them at the right temperature and protecting them from predators. Parents often have to work extremely hard to feed their infants. During daylight, many small birds feed their young four to twelve times an hour on average. With several nestlings to feed, the parents will be feeding them hundreds of meals during the course of a day. Mammalian mothers face the additional problem that only a fraction of the energy in the food they eat is available to the young through their milk. Consequently, lactating mothers have to eat a good deal more than usual; lactating Bank voles, for example, eat nearly twice the amount consumed by females that are not pregnant or lactating. The amount of milk that mothers can produce is impressive. Mother Grey seals suckle their single pups for about 17 or 18 days from birth. Over this period, the pups roughly treble their weight to about 50kg (110lb), becoming barrel shaped in the process. During this time the mother does not feed and loses a lot of weight.

Among mammals, it is usually the mothers that provide much of the care for the young but this is not always the case. Fathers (marmosets, for example), older brothers and sisters, sisters of the mother (Rhesus monkeys), and even completely unrelated individuals can play a part, depending on the species. A successful parent is one that, during the course of its life, raises a large number of offspring which themselves go on to

► **The first bond.** Immediately after birth a ewe licks the lamb and so becomes familiar with its taste and smell.

◄ **Encouraging independence** in her offspring, a mother baboon (1) rises and withdraws contact with her infant, (2) moves a few steps with the infant following and (3) turns to reassure the infant who stops. Older infants which have gained some independence are rejected more often (4).

Bonds between Parents and Young

Among the animals in which parents continue to care for their young for much of their infancy, there is a great deal of variation in how the infants stand up to being separated from their parents for brief periods. In some species, such separations are perfectly routine while in others they are highly disturbing to the young. Much depends on the way of life of the animals concerned. Young birds that are relatively helpless as nestlings, such as starlings or blackbirds, are frequently left on their own as their parents go out to forage for food, and remain quietly there until the parents return.

By contrast, birds that are active and mobile at birth, such as goslings and ducklings, rely for their food and protection on their mother and, if they are separated from her or from one another, they start to give loud calls which often hasten the mother's return. Similarly, mouse and rat pups, which are naked, immobile and blind at birth, show no reaction to the departure of the mother from the nest, whereas baby monkeys, which are carried by their mother nearly all of the time in the first days of life, cry with distress if separated from her. These cries have a powerful effect on the parent

and she usually picks up her baby very rapidly.

In many kinds of animals, including sheep, goats, antelopes, and other hoofed mammals, and in many primates, it is clear that a strong bond develops between the mother and her young. If this bond is threatened, by separation or by some other disturbance, the baby will behave in a way that is likely to reestablish it, and if this fails, the baby's behavior will be deeply disturbed. Baby monkeys and apes separated from their mothers for any length of time tend to become immobile and depressed, and

may die, even in some cases when other members of the group try to adopt the orphan. This shows that the bond that forms is a quite specific one between a particular mother and her infant, rather than that the mother is prepared to care for any infant, or that the infant is willing to accept care from any adult. In hoofed mammals, such as sheep, this bonding process starts very early in life. Once she has given birth, the mother licks the baby, and so tastes the birth fluids on the lamb's body. Once she has learnt the smell and taste, she will not accept a lamb with a different smell or taste.

become parents. Less successful parents raise fewer offspring.

There are some things affecting a parent's success over which it may have rather little control: a drought or epidemic may kill all its young irrespective of any action it takes. Other things, however, may be under the parent's control. In particular, a mother may be able to vary the number of young produced in any one batch: the number of eggs in the clutch she lays or the number of pups in her litter. Female swifts, for example, usually lay three eggs in a clutch but may produce fewer. Why do they not lay more eggs? Would a mother that habitually produces large clutches be more successful in the course of her life than one that produces smaller clutches? Most probably she would not. Female swifts that were given artificially large broods to raise had much greater difficulty finding enough food for their young than did females with broods of only two or three nestlings. As a result, many of the young birds from the larger clutches died. It turned out that mothers that laid clutches of three eggs on average successfully reared more young than those that laid either smaller or larger clutches.

Second, a mother may be able to control the length of time for which she continues to offer care to her young in a particular brood or litter, before abandoning them, mating again, and starting another brood or litter. If she abandons her present set of young too early, they may not survive, or at the very least, the chances of some of them surviving may be reduced. On the other hand, if she remains with them after the time when they can fend for themselves, she has less of her own lifetime left in which to produce more young. The mothers that do best will be those that leave their current set of young at the optimum time. This is not to say that the young of her current batch will also do best if the mother leaves them at this particular point: they would do better if their mothers stayed with them for longer rather than going off and raising other young. As a result, there will come a time when the young demand care from their parents but they are no longer willing to provide it. Many animals, show this type of conflict between parents and offspring around the time of weaning.

Many young animals are virtually helpless shortly after birth or hatching—they are said to be altricial; they can do little more than take in the food that the parents provide. Baby thrushes can gape with their bills when the parents bring food to the nest; a baby chimpanzee can suck from its mother's nipple and move its head from side to side if its mouth is not in contact with it, but the mother has to hold on to her young in case they fall off, and she must clean them, carry them and protect them. As the young animals grow older, however, so their abilities develop, and the parents become less attentive. The young start to run, walk, fly or swim more competently and begin to feed for themselves so that the parents become less protective. To a lesser extent, parental care also wanes among precocial species where the young are mobile and active from the start of their lives.

What makes very young animals attractive to their parents in the first place? And what causes parental care to diminish as the baby grows older? Very young animals possess features that are known to be very appealing to parents. When they are cold, baby mice give off ultrasonic calls, that is, squeaks that are so high pitched that they are beyond the range of human hearing. This often happens when they stray from the protective warmth of the nest, and the mother responds by going to the pups, picking them up in her mouth and taking them back. In comparison with the adults of the species, the faces of many young birds and mammals are more rounded and have bigger eyes. This gives them a "cute" appearance to us, and it is this combination of features that seems to make them so attractive to their parents. Many baby monkeys are a different color from the adults. Baby baboons are black and pink, for example, whereas the adults are olive. These babies are not only highly attractive to their mothers but to other members of their social group, and adult and adolescent females are often seen trying to steal newborn babies from the

▲ **Caring spider.** A female wolf spider (*Pardosa* species) with young on her back. Wolf spiders are vagrants on open ground and they carry around the egg cases as well until the young hatch.

◀ **Bursting forth** from their mother's back, froglets emerge from a female marsupial frog (*Gastrotheca ovifera*). The mother has a brood sac on her back in which the eggs develop through all stages of metamorphosis.

◄ **Helpless offspring,** such as those of the Brown rat (*Rattus norvegicus*), are known as altricial young. The mother must take care of all their needs in early life.

▼ **Eggs and parents.** Strategies of parental investment and care in the production of eggs and young. (1) The perch shows no parental care after eggs are laid and most are either eaten by predators, fail to hatch or hatchlings fail to survive— if one egg survives to adulthood that is a good average. To compensate for this the female perch weighing 1kg (2.2lb) lays up to 200,000 eggs, each no more than 1mm in diameter. (2) In the kiwi, both parents contribute to the raising of the chicks, which have a high probability of surviving. The female weighing 1.8kg (4lb) therefore lays just one enormous egg up to 13cm (5in) long and weighing 0.45kg (1lb) which in relation to body size of female, makes this the largest egg of any bird.

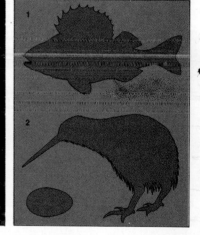

mother. This behavior vanishes as the babies lose their black and pink coloring and take on the adult color.

As babies grow older, they become more independent and capable of looking after themselves, and the parents actively promote this independence. Even at two weeks of age, baby baboons will be encouraged towards independence by their mothers by playing games with them in which they place the babies on the ground, walk away a few paces from them and then turn back giving them a gesture of reassurance. Later, the mothers will tend to leave their babies more often, and will also often reject the babies' attempts to suckle. This process of promoting the infant's independence is called "weaning." In monkeys and apes, it is often very gradual, and mothers will cuddle their young long after they have stopped feeding them. Indeed, the main spur to independence appears to be the mother's resumption of sexual activity, when she may begin to tolerate her young far less than she did before. In chimpanzees, the infant's psychological dependence on its mother may last for years. Indeed, two- to three-year-old

chimpanzees that become orphans may die within weeks of their mother's death because of depression.

Parental care declines as the young grow older. But it also varies for other reasons. One important factor is the parents' experience. Mother mice that have raised previous litters respond more readily to their pups than mothers with their first litters; Rhesus monkeys are less adept at looking after their first-born than after later infants. In marmosets, which are small South American monkeys, juveniles that have looked after their brothers and sisters make better parents than those that have not. The kind of parental care a young animal receives can also depend upon its sex. In Bonnet and Rhesus macaques, adults are more aggressive to mothers with daughters than to mothers with sons and, as a result, mothers are more protective towards their daughters than to their sons. Strangely enough, in another species of macaque, the Pigtail macaque, adults are more aggressive to females that are pregnant with female fetuses than to females bearing males. The reason for this, is not known.   NRC

BREEDING SYSTEMS

Stable pairs. . . Single males and many females—harems. . .
The role of females in reproduction. . . The male's role. . .
Breeding in elephant seals. . . Difference in size between males
and females. . . Choosing a mate. . . The sex roles in the
Spotted sandpiper. . . Satellite males—stealing the show. . .

WHAT do foxes, gibbons, beavers, elephant shrews and the tiny antelope, the dik-dik, have in common? These five mammals, representing five different major groups, all live in stable monogamous pairs, whereas most mammals do not. A much more common mating system is that in which some males monopolize several females each, while other males are barred from breeding. During the Red deer's rut, for example, the most aggressively dominant stags guard "harems" of up to a dozen fertile hinds while lesser stags are consigned to bachelor groups.

"Polygyny" (multiple female) is the term used to denote a mating system, such as the Red deer's, in which a single male may sire offspring by several females. "Polyandry" (multiple male) is the converse: the system in which a single female may breed with several males. Polyandrous mating systems are extremely rare in mammals—suspected in a handful of species and well documented in none—while polygynous systems are numerous. Indeed, polygyny is much more common than polyandry in the animal kingdom generally. To see why this should be so, it is necessary to consider the most fundamental distinction between the sexes.

The female is, by definition, the sex that produces the larger sex cell: an egg cell is larger than a sperm. This size difference has some immense consequences. By producing relatively large eggs, females initially put more energy into each offspring than males do. In those animals that simply lay their eggs in a suitable environment and leave them to develop, a female's reproductive output is limited by her capacity to garner energy and produce eggs. Her mate, however, achieves the same reproductive success for a far smaller energetic expenditure. It follows that the opportunity to fertilize a female's eggs is a contested prize. Males compete for access to females and a successful male may sire more young than any one female could ever produce. It is because the male makes a lesser "parental investment" in each offspring that he has a higher reproductive potential than the female and can enhance his reproductive output if he can mate with many females. Female fitness derives less profit from mating with many males. One male is all it takes to fertilize all the female's eggs.

Of course, many animals do not simply lay their eggs and abandon them. Nevertheless, it remains typical for females to invest more in each offspring than males. Due to the larger egg's relative immobility, fertilization within the maternal body has evolved many times, within the paternal body almost never. Internal fertilization then set the stage for several evolutionary innovations that magnified the initial sex difference in parental investment; pregnancy and lactation in mammals are particularly dramatic examples. In many large mammals, for example, the female's reproductive capacity is physiologically limited to the rearing of a single offspring every year or two. Access to

▲▶ **Sexual dimorphism and type of breeding system.** (1) and (2) Monogamous breeders tend not to be dimorphic in size. (3), (4), (5) Polygynous breeders that are dimorphic. (1) Common tern (*Sterna hirundo*) with male bringing food during courtship to female, who is sitting on the selected nest site; terns generally pair for life. (2) Moloch gibbons (*Hylobates moloch*) in a mutual duet which not only helps in forming the pair bond, but in maintaining and developing it. (3) Male (larger) and female Northern elephant seal (*Mirounga angustirostris*) on mating beach. Each season only the highest status males maintain a harem and are successful breeders. (4) Male (larger) and female Black howler monkeys (*Alouatta caraya*) howling; males may monopolize up to three females. (5) Male capercaillie (*Tetrao urogallus*) calling on courting ground (lek), with drab female who is in a mating invitation position.

all the males in the world would not change that limit. The male's reproductive output, on the other hand, is a direct function of the number of mates he can acquire. Natural selection favors those attributes that maximize individual reproductive success, so that it is little wonder that males so often vie for polygynous status.

The Northern elephant seal, breeding on Pacific islands, is the most extremely polygynous mammal known. In early December, the huge bulls haul out on to the beaches, where they remain without feeding until March. There they battle with one another, at great energetic cost and with occasional severe wounding, until the winners have established dominion over certain stretches of beach. Pregnant cows begin to arrive in mid-December. Each delivers her pup on the beach and then nurses it for about four weeks before departing. Shortly before that departure, the moment arrives that makes the bulls' struggles worthwhile: she mates and conceives the next season's pup. There is enormous variability in the mating success of the

2

3

5

bulls. Only about 10 percent of the male pups born survive to join the breeding competition at five or six years of age, and most of these never copulate. About one in 100 males lives to be nine or ten years old, which is the age of the most successful bulls. A rare individual will then sire 100 or even 200 pups, whereas the most successful of cows might bear 10 in a lifetime. When the season is over, the males are debilitated. Some will have lost as much as a third of their December body weight. Many die very shortly after the ordeal of the beach.

The dense aggregation of females permits the extreme polygyny of elephant seals. Most animal mating situations have a lesser "potential for polygyny" because, typically, females spread themselves out to avoid costly competition for food or other resources. Mating systems vary, therefore, from extreme polygyny like this. through lesser degrees of it, to monogamy and to polyandry. Monogamy is commonly associated with widely dispersed females: each couple occupies a territory from

which both individuals are concerned to exclude competitors of their own sex.

In a monogamous mating system, the sexes generally share parental duties, and the male tends not to be noticeably larger than the female. In highly polygynous animals, by contrast, males play no parental role, channeling their energy into competition for mates. The demands of that competition have often led to the evolution of large male size and well-developed armaments. If we compare related species within the seals or the primates, for example, the degree of "sexual dimorphism" (the difference in male versus female body size) which characterizes the species is closely correlated with the degree of polygyny. In the intensely polygynous elephant seal, one bull weighs as much as three or four mature cows. This relationship is not confined to mammals. Most birds, for example, are monogamous, with the sexes of equal size but, in the few highly polygynous species, such as the capercaillie and some other grouse,

◄ **Satellite males—stealing the show.** In intensely polygynous mating systems where competition between males is fierce and costly, it is often found that some males opt out of the majority mode of competition. For example most Golden toads (*Bufo periglenes*) attract females by their orange coloration and by sitting conspicuously on a pool edge. Other "satellite" males, however, wait further away, hardly contributing to the display at all, but are able to intercept females approaching the main display area. In other frog species, calling attracts females, but satellite males are silent.

Just the same sort of satellite behavior has been described in crickets. Singing males attract females, but they also attract warbleflies that lay their eggs on the singers and thereby eventually kill them. The silent satellites do not get as many females as the singers but they are not parasitized either.

▼ **Lone egg layer.** In many species, such as this Leatherback sea turtle (*Dermochelys coriacea*) females lay many eggs and the hatchlings have to fend for themselves, with a high resulting mortality.

The Sex Roles in the Spotted Sandpiper

More than 90 percent of all bird species typically breed in monogamous pairs, and those that do not are mostly polygynous or promiscuous. But a few species breed in stable polyandrous associations of one female and her several males.

One such species is the Spotted sandpiper, a common North American wader. The females, substantially larger than the males, arrive first at the breeding grounds and compete aggressively for territories. The female is also the active partner in courtship and, once she succeeds in attracting a male, she sets him up with a clutch of four eggs and leaves him to care for them while she goes courting again.

If, as often happens, a nest is destroyed by a predator, the female will return to provide her mate with replacement eggs and, if she does not succeed in attracting another male, she will help out a bit during incubation. A female's senior mate may resist the addition of another, attacking the new male but, if so, the female soon puts him in his place. Her mates then remain mutually hostile.

In all of these features, the Spotted sandpiper reverses the sex roles characteristic of polygynous birds. Access to males has become the limiting resource for female reproduction, rather than the reverse as is true of most animals. A male simply does not have the time to raise more than a single brood of four in a season, whereas successful females may fledge young by two or more males, and have been known to fledge up to nine chicks in one summer. While the males are sitting on eggs and guarding chicks, the females are able to channel time and energy into foraging and egg production.

the males are usually much larger than females of the species.

A female commits a lot of time or energy to each individual offspring, so that she must choose her mate with care. Mating with a male of the wrong species can lead to months of wasted effort rearing an inviable or infertile hybrid. Mating with a genetically inferior male of her own species is not much better. The male, on the other hand, with his lesser parental investment, has less need to be choosey. He often commits little more than a few sperm and sperm are cheap. So it is common for males to court rather indiscriminately, and for females to be selective. A male hamster or guppy, for example, will initially court females of related species as readily as his own, whereas even inexperienced females spurn inappropriate males. If your galoshes have ever been seized by an amorous frog, you know about indiscriminate males!

The exercise of mate choice gives females some leverage. It is precisely because females are a "limiting resource" for which males compete that females are often in a position to demand something from courting males. In many insects, for example, males transfer valuable protein to females along with their sperm in the form of spermatophores that may amount to a quarter of the male's body weight. In predatory animals, such as fish-eating birds, females may make mating conditional upon the provision of prey. A male tern, for example, passes the female a fish precisely at the moment of copulatory contact. The result is that the energetic cost of egg production is split between the sexes. Wherever the male's parental investment approaches that of the female, the time he can devote to polygynous competition is reduced, and we are apt to observe instead a stable monogamy.   MD

The Origins of Behavior

How do animals come to do the particular things that they do? This question used to be phrased differently to ask: "Is it instinctive or is it learned?" But, as behavior has been studied in more detail, ethologists have realized that behavior cannot simply be split up that way and that its development involves a lot of subtleties. An animal's inheritance affects its ability to learn so it may learn some things much more easily than others; at the other extreme, there are many behavior patterns which develop without learning but this does not mean that they are inevitable or "blueprinted in the genes." Even very fixed behavior patterns can be altered by a change in environment: a hopeful message when we come to think of some of the less desirable features of our own species. Behavior has evolved so that animals behave in an adaptive fashion at each stage of their lives. To achieve this, all the behavior shown by animals develops through a subtle interplay between their inheritance and the environment in which they find themselves. The articles in this section of the book show just how subtle are the interrelationships between these factors.

◄ **Duet in the tree tops**—a pair of grackles (*Quiscalus* species) singing.

LEARNING

Negative learning—habituation. . . Trial and error or instrumental learning. . . Thorndike and Skinner box tests. . . Learning in jackdaws. . . Avoiding unpleasant experiences. . . Learning to identify individuals. . . Finding out about their surroundings. . . What is intelligence?. . . Memory in animals. . . Can animals count?. . . Insight. . . Bird song learning. . .

From birth to adulthood, a human assimilates a vast array of information. He or she learns to speak the language of his or her native country, to read and write, to ride a bicycle, to drive a car, or to operate a computer. A domestic cat will learn that, to gain access to its home and to be fed, it must open a cat flap and appear at the correct time of day in the right house. Similarly, a foraging bird will learn that, if it is not to be stung or poisoned, it should avoid warningly colored wasps or bees. The behavior patterns which have been passed on from generation to generation are modified and developed by learning about the environment in which the animal finds itself.

To behave appropriately, an animal must possess information about the correct way in which to perform a range of complex behavior patterns as well as about the correct circumstances in which to perform each one. For example, as well as "knowing" how to go about hunting, catching, killing and dismembering a mouse, a hawk must also "know" that a mouse is the right sort of animal to hunt and that hunting is the right sort of behavior to engage in when feeling hungry. Of course, this knowledge may not be conscious; nevertheless, all behavior relies for its success on the animal possessing certain kinds of information about itself and about the world in which it lives.

Learning is one way in which this information can be obtained. Learning takes many forms and is more important to some species than to others, but all animals seem capable of learning to at least some extent. In humans, learning has taken on a degree of importance unique in the animal kingdom; almost everything we do, from brushing our teeth in the morning to flicking off the light switch at night, has been learned. But, while no other species surpasses us in the sheer amount and range of information that it acquires through learning, there are many examples of animals learning things that we would find difficult or impossible. For example, birds learn the detailed structure of songs which are far too rapid for us to follow, and salmon home in on the precise chemical characteristics of the stream in which they hatched years earlier.

Of the various types of learning that have been identified, the simplest and probably the most common is the process known as "habituation." Habituation is, in a way, a kind of negative learning process: it consists of learning that a particular event in the environment is of no consequence and can therefore be ignored. Imagine what happens if your pet dog is asleep at your feet and you suddenly tap loudly on the table—the chances are that the dog will wake up, prick up its ears, and perhaps get up to see what is going on. A few minutes later, when the dog has settled down again, you tap again—this

Unnamed Numbers: Can Animals Count?

At the beginning of the twentieth century a German mathematician, von Osten, caused a sensation by announcing that he had taught a horse to count. When von Osten called out a number, the horse would tap the same number of times with its hoof. The horse, appropriately, was called "Clever Hans."

Unfortunately for von Osten, a psychologist named Pfungst revealed that Clever Hans was not really able to count at all. Instead the horse was "cheating" by watching for small involuntary movements by von Osten. For example, when von Osten called out the number six, Hans would start to tap on the ground with his hoof. At the sixth tap, von Osten, anxious that Hans would give the right answer, would unconsciously give the game away by holding his breath or adopting a slightly more attentive posture. Seeing these minute movements on the part of his trainer, Hans would stop tapping. Pfungst proved his case by showing that Hans was unable to give a correct answer if the requisite number was not known to his trainer, or if Hans was prevented from seeing his trainer. The case of Clever Hans is important because it shows how careful we must be when assessing the learning abilities of animals.

Is there, then, any real evidence that animals can count? Otto Koehler devised a task in which a bird (pigeon, jackdaw, raven or parrot) was shown a sign on which were painted a number of dots, followed by a selection of food dishes with different numbers of dots on their lids. The bird's task was to select the food dish with the same number of dots as had appeared on the sign. For example, if the sign showed five dots, the bird had to choose the dish which had five dots on its lid. To prevent the possibility of cheating, the bird was separated from its trainer by a screen, and the precise pattern of dots on the sign and on the dishes, as well as the number of dots, was changed between trials.

By means of such careful and painstaking experiments, Koehler proved reasonably conclusively that his birds could count. However, the counting abilities of the birds were mediocre by human standards: a parrot named Geier and a raven called Jacob were able to count up to six after prolonged training, but this was the limit of their achievement.

► **Venturing out to play,** two young Eurasian badgers (*Meles meles*) emerge from their sett. During the coming weeks they will learn progressively more about their surroundings until they achieve independence from their parents.

▼ **Learning to live together.** In social animals such as Hamadryas baboons, each individual must discover its place in the interlocking web of relationships. This is achieved progressively during activities such as those shown here. (**1**) Infants playing. (**2**) Juvenile exploring. (**3**) Female presenting to male (not shown) during courtship). (**4**) Female grooming a male, (**5**) Foraging as part of a group. (**6**) Aggressive encounter between males.

2

5

6

time the dog might prick up its ears but not bother to get up and investigate. Repeat the tap a few more times and the dog will cease to take any interest in it at all, and will go on sleeping peacefully. This is habituation: the dog has learned that a tap on the table does not signify anything important, such as the presence of danger or the arrival of food, and that it can therefore safely be ignored.

Habituation has been demonstrated in many species of animals, from the simple amoeba to humans. It may even be a property of all living cells. It is a vital process because, without it, the animal, bombarded by sights, sounds, smells and touches, would be in a constant and mostly needless state of alarm or expectation. Habituation filters out the multitude of background stimuli that have no importance and, in doing so, leaves the animal's attention free to concentrate on what really is important.

This leaves us with the question of how animals learn about events that are positively useful to them, such as the location of food or water. One way is through the type of learning known as "trial and error" or "instrumental" learning, in which the animal learns that, by making a particular response, it obtains a reward. Instrumental learning was first analyzed by the American psychologist, E.L. Thorndike, at around the turn of the century, using an apparatus called a "puzzle box." The puzzle box was simply a wooden cage which could be opened

from the inside by pressing down on a foot pedal. A cat was placed in the box and left to discover for itself how to get out. Over a series of trials the cat would gradually learn that, to get out, it must press down the foot pedal, which in turn would cause the door to open. Thorndike coined the term "instrumental learning" because the response (in this case, pressing a pedal) is instrumental in leading to the reward (that is, being let out of the puzzle box).

Ever since Thorndike's first demonstration of the phenomenon, instrumental learning has been a favorite subject of psychologists. Nowadays, however, they usually study rats rather than cats, and the apparatus is a maze or a Skinner box rather than Thorndike's puzzle box. In a maze, the rat learns which way it must turn at each intersection if it is to obtain a reward of food; in a Skinner box, it learns to press a lever to obtain food. Instrumental learning is also called "trial-and-error learning," because the animal must discover for itself what response it must make to obtain the reward. In a Skinner-box task, however, the psychologist usually helps the rat to learn by a process known as "shaping." First, the rat is rewarded with food every time it approaches the lever; then it is rewarded only when it places a paw on the lever; finally, it is rewarded only when it presses down the lever.

Instrumental learning is most easily studied in the laboratory, but it plays an important part in the everyday lives of

---

Memory in Animals

Memory is not the same as learning but it is an essential part of learning. The term "learning" usually refers to the process whereby an animal acquires information from the environment, whereas "memory" usually means the process whereby that information is stored in the brain, ready for future use. Without memory the process of learning would be useless because the learned information would immediately be forgotten.

While the saying that "elephants never forget" is probably just an old wives' tale, it is well known that some animals have especially good memories for certain sorts of information. As has been mentioned earlier (see p16) birds, such as Marsh tits and European jays, and mammals, such as Red and Gray squirrels, hide away seeds and nuts when food is plentiful and then return to their hoards when food is scarce. This means that the location of stored food must be remembered and, in some cases, this involves remembering more than 100 separate,

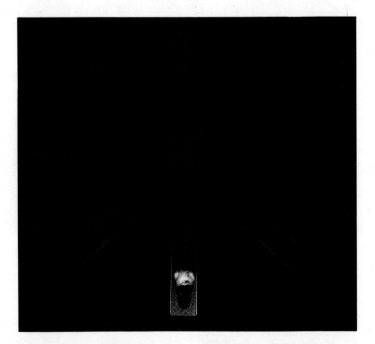

individual hiding places.

David Olton has tried to measure memory for spatial locations in the rat, using a specially invented eight-armed "radial maze," ABOVE Food is placed at the end of each arm of the maze, and the rat has to learn how to collect all of the

food in the most efficient manner. Obviously, the most efficient strategy is for the rat to go just once to each of the eight arms, and rats prove very adept at learning this strategy. This means that, as it is making its way around the maze, the rat must remember which arms it

has already visited so that it can avoid visiting them again.

How exactly does the rat know which arm is which when all the arms look so similar? One possibility is that the rat might just go round the arms in sequence, say, by going from each one to the next in a clockwise direction. But, if a record is kept of the rat's movement, it is found that the arms are not visited in any consistent sequence. Alternatively, the rat might leave a scent trail as it goes from arm to arm, so that arms that have already been visited could be identified by smell. Again, however, careful experiments have ruled out this possibility.

The answer seems to be that the rat identifies each arm in relation to landmarks in the surrounding environment. It learns, for example, that one particular arm is near a window while another is underneath a lampshade. If the relevant cues are moved around the rat becomes confused, and is unable to remember which arms it has already visited.

▶ **Inexperienced albatrosses.** In an experiment with Laysan albatrosses (*Diomedea immutabilis*) birds with little breeding experience will quite happily incubate a grapefruit! They do not seem to know what an egg should look like.

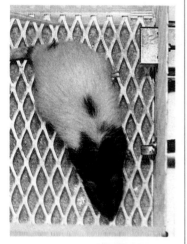

◀ **Instrumental learning**—the behavior of a rat in a Skinner box. The rat begins by exploring the box at random TOP. When it presses the lever CENTER a food pellet is automatically delivered to the food-cup and is eaten by the rat BOTTOM. Soon the rat learns that by pressing the lever repeatedly it will get a continuous supply of food, and the response becomes automatic and rapid. Instrumental learning is then complete.

◀ **Rat in a radial maze** FAR LEFT. The complete path taken by a rat in the process of collecting food from each arm of the maze is shown: the picture was taken by attaching a small red light to the rat's neck. The main points to note are that the rat only enters each arm of the maze once, and that it does not adopt a simple strategy such as following round from one arm to the next in a clockwise direction (see box).

many animals, and of humans. When a young jackdaw starts to build its first nest, it will collect and attempt to incorporate into the structure almost anything it can find—bits of paper or glass, empty cans, old light bulbs. It soon discovers, however, that particular kinds of twigs and grasses make very good nest material whereas stones and cans do not! Whenever an animal modifies its behavior in the course of obtaining a reward, such as food, water, a mate or a nest, we can be fairly sure that instrumental learning is responsible.

Instrumental learning usually involves discovering how to find or deal with some kind of desirable reward but, for many animals, it is just as important to know what kinds of events in the environment are potentially unpleasant or dangerous. Here we meet yet another category of learning, usually known as "avoidance learning." One of the best examples of this is to be found in the behavior of rats towards poisoned bait. Rats tend to be suspicious of any new food that they encounter so that, if a rat finds poisoned bait in its territory, it is likely at first to eat only a very small amount. This means that, unless the poison is very powerful, the rat will be made ill but will not have eaten enough poison to be killed. Once poisoned in this way, a rat will never eat the same bait again; it has learned to associate the bait with the unpleasant symptoms of illness and, as a result, it avoids the bait thereafter. Avoidance learning of this type is dramatic for its speed and its permanence, in that a single experience of poisoning will put a rat off bait for the rest of its life. This type of learning is, in fact, one of

Insight

When someone sits down to think about a problem and then suddenly "sees" the correct solution, we say that he or she has reached the solution by a flash of "insight." Because insight is a purely mental process, it is usually regarded as the highest form of learning. Do animals use insight in the same way as humans?

Some of the best-known experiments on insight were those carried out by Wolfgang Köhler early this century, on captive chimpanzees. The chimpanzees had to work out how to obtain bunches of bananas that were placed out of direct reach. The solution to this problem was to fit together a number of sticks so as to form a long pole, or to stack empty boxes on top of one another and then climb up to within reach of the bananas.

Köhler reported that the chimps did show evidence of insight—when faced with the problem of reaching the bananas they sat around for some time apparently contemplating the sticks, boxes and bananas, and then they suddenly set about fitting the sticks together or piling up the boxes in an appropriate way. Köhler concluded that the chimps could successfully appreciate a problem and solve it by purely mental processes. Others, however, have pointed out that chimps will fit sticks together and climb up on piles of boxes even when there is nothing for them to reach: apparently it is just in the nature of chimps to play with sticks and boxes in this manner. Thus, it is unclear whether

Köhler's chimps were engaging in genuinely insightful behavior, or whether they just happened to solve the problem unintentionally in play.

Chimpanzees in the wild also use tools to get food. One favorite trick is to "fish" for termites by poking a thin twig into a termite mound. When disturbed in this way, the termites cling on to the twig so that, if the twig is carefully pulled out of the mound, the termites can be eaten (ABOVE). This might be insight but, in fact, young chimpanzees seem to learn the technique by a combination of imitation and trial-and-error.

If insight is so difficult to demonstrate in animals, should we doubt its existence in humans? No-one would deny that humans can solve problems purely by thinking about them in the abstract, but a great deal of what we call insight may be a kind of mental process of trial-and-error. As such, insight may not be all that different from simpler kinds of learning.

▲ **Tool-using chimp**—a young chimpanzee (*Pan troglodytes*) uses a stick to extract termites from a nest.

▶ **Deadly play**—a thin line divides play from real life. This young cheetah first played with the Thomson's gazelle (*Gazella thomsoni*) fawn, but the cub had learned enough in its short life to realize that its playmate was also food, and finally killed it.

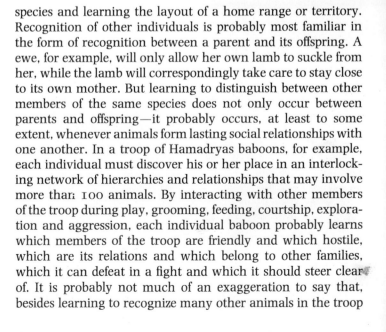

the major reasons why rats have so far survived all attempts to eradicate them.

Other animals also learn to avoid foods which taste unpleasant or which make them ill. Conversely, animals will come to prefer and to seek out foods that taste especially pleasant. For example, if a young chick is given a mealworm, it will initially show signs of alarm. Eventually, however, it will probably peck at the mealworm out of curiosity, and will discover that mealworms taste good. Once this discovery has been made the chick will forget its original fear, and will enthusiastically consume any mealworms that it finds.

Two more forms of learning that are common among animals are learning to recognize other individuals of the same species and learning the layout of a home range or territory. Recognition of other individuals is probably most familiar in the form of recognition between a parent and its offspring. A ewe, for example, will only allow her own lamb to suckle from her, while the lamb will correspondingly take care to stay close to its own mother. But learning to distinguish between other members of the same species does not only occur between parents and offspring—it probably occurs, at least to some extent, whenever animals form lasting social relationships with one another. In a troop of Hamadryas baboons, for example, each individual must discover his or her place in an interlocking network of hierarchies and relationships that may involve more than 100 animals. By interacting with other members of the troop during play, grooming, feeding, courtship, exploration and aggression, each individual baboon probably learns which members of the troop are friendly and which hostile, which are its relations and which belong to other families, which it can defeat in a fight and which it should steer clear of. It is probably not much of an exaggeration to say that, besides learning to recognize many other animals in the troop

as individuals, the baboon is also assimilating knowledge of the personalities, life histories and relationships of those individuals.

The ability to learn the layout of a home range or territory is known as "spatial learning." This is something that is essential if the animal is to be able to locate sources of food or water, or find its way back to a nest or den, without having to search at random each time. Given the importance of spatial learning, it is not surprising that the first thing that almost any animal will do, if it finds itself in a new environment, is to embark on a thorough program of exploration in which every nook and cranny is investigated. During this exploration, the animal learns where important things, such as food and water, are located, and also how to move most efficiently from one place to another. In short, the animal builds up a kind of mental "map" of its territory, on which landmarks and points of interest are marked, a phenomenon which has been demonstrated in species as diverse as bees, homing pigeons, mice and foxes.

To summarize, animals have been shown to be capable of learning to ignore irrelevant events in the environment, to perform artificial responses such as pressing a lever when these are rewarded with food, to build structures such as nests, to avoid eating things that taste nasty or that make them ill, to recognize other individuals and to find their way around. Does this mean that animals are intelligent? Scientists who study learning cannot, unfortunately, agree on a common definition of intelligence. People who ride horses often say that horses are intelligent because they understand the rider's commands but it could equally be argued that, if horses were intelligent, they would not allow themselves to be ridden in the first place! Thus, whether or not an animal is considered intelligent depends upon what is meant by intelligence. Certain species of animals are capable of feats of learning that would be considered prodigious even by human standards, but usually these are rather specialized kinds of learning abilities. "Intelligence" is usually thought of as a general ability to learn, and only in humans has learning transcended its biological limitations to an extent that it can be applied to almost any problem which the environment presents.   TJR

Copycat or Composer?
Bird song learning

► **Sonagrams of chaffinch songs.**
(1) Three songs recorded from different chaffinches in the Orkney islands, shown here as sonagrams (plots of frequency against time which give a visual image of the sound). Each of these three songs consists of three successive phrases, within which the notes are nearly identical, followed by a longer rising and falling syllable, the end phrase or flourish. These three illustrate well how different birds may sing very similar songs, a result of the fact that they copy from each other, often with great accuracy.
(2) Experiments on song learning. Traces (**a**) and (**c**) are sonagrams of song phrases played to young birds during their first year of life, and the (**b**) and (**d**) traces are the songs that those young birds produced. In both cases the copying is very accurate. In (**b**) the young bird was played the song in its first spring, when it was starting to sing itself. In (**d**) the bird was trained when it was a fledgling, during its first two months of life, long before it sang itself. Remarkably, such young birds can store the memory of songs they have heard over their first winter and then sing accurate copies of them many months later.

◄ **Proclaiming a territory held,**
a Savanna sparrow (*Passerculus sandwichensis*) sings from a prominent perch. In most birds only the male sings and males with more varied songs may be more successful in getting mates because females find them more attractive.

It is well known that some birds can imitate sounds that they hear. For example, budgerigars can be taught to say "Hello" and Hill mynahs to imitate the closing of doors. Parrots and mynahs can build up quite a large vocabulary of different sounds, all copied from people or from noises they hear frequently around their cages rather than from other birds of their own species. Occasionally, it may be obvious that birds in the wild also imitate sounds; for example, a thrush may produce a trill identical with that of a telephone. This is not often so, however, and most of the sounds that birds make are easy to identify as coming from their own species rather than being copied from elsewhere. But this does not mean that they do not imitate: it simply means that such imitation as there is, normally takes place from other members of their species.

One of the first detailed studies of how the songs of young birds develop was carried out about 30 years ago in England by Professor W.H. Thorpe of Cambridge University. He studied the songs of male chaffinches which he reared by hand from a few days after they hatched. To analyze the songs that they produced, he used the sound spectrograph, a machine which had only recently been developed; it produces a chart of the pattern of the song which can be studied visually.

The most striking effect he found was if a young bird was kept in isolation so that it could not hear other birds. In this case, the song that it produced when it started to sing as a young adult was very different from normal chaffinch song. It was about the right length and in the correct frequency range, but it lacked the "terminal flourish" with which chaffinches always sign off at the end of their songs. It was also not clearly split up into separate phrases, the successive notes being rather more simple and similar to one another than in normal song. This experiment suggested that young chaffinches would only sing normally if they were not kept alone: but was this because they usually copy their songs from others? By playing recordings of songs to other young birds and finding that they produced precise copies of what they had heard, Thorpe showed that this was the case. He even tutored one bird on a recording which he had edited so that the flourish appeared in the middle rather than at the end, and the young one copied it although no wild bird sings in that way.

Chaffinches normally only sing chaffinch song, although they can be trained to sing phrases from certain other birds which have songs very like their own, such as the canary and

Tree pipit. The development of their song is a fine example of how learning and inheritance both contribute to the development of behavior in a very complicated way. Chaffinches clearly learn the songs that they sing, and usually the copying is very precise, the finest details being reproduced faithfully. But they will not learn just anything. They seem to hatch with a rough idea of what chaffinch song should sound like and will only learn to sing patterns which match this. So, although the song of many other bird species can be heard where chaffinches nest, they learn that of their own species.

Since Thorpe's work, the songs of many other songbird species have been studied and, although copying from others has been found to be a feature of song development in all of them, other aspects have been found to differ among them. For example, while young chaffinches only learn songs in the few months between hatching and first singing themselves, some other birds may go on learning throughout life. The canary is a good example here; in this species, males drop some phrases and develop other new ones each year. Another difference is in the accuracy of copying. Most chaffinches learn their songs very accurately so that many of the males in a wood may sing very similarly to one another. This is rather like people copying their accents from one another and, just as people in London sound different from people in Liverpool, so chaffinch song varies from one place to another as well. Some other species show much less accuracy and more improvization. In these a great variety of different songs may be heard in the same area, and regional dialects are not so easy to detect.

Improvization is one way that males can develop an interesting and varied song with many different phrases. Another way of achieving this is not by improvization but by accurate copying of a wide variety of sounds including those of other species. Chaffinches in the wild do not do this and have rather simple songs which are probably used mainly in communication between males on neighboring territories rather than in mate attraction. Some other species can, however, develop enormous repertoires of different songs by copying almost any sound they hear which they are capable of reproducing. The Mocking bird of North America is famous for the variety of species of birds it will imitate, but the prize probably goes to the European Marsh warbler in which each male imitates, on average, 76 other species, from the African countries to which it migrates as well as from its nesting area.   PJBS

INHERITANCE AND BEHAVIOR

Is behavior instinctive or learned?—nature or nurture...
Copying language... Learning to feed in gulls... How much
do genes affect behavior in mice?... Selection of behavior by
artificial breeding... The inheritance of learning... Do
behavior patterns develop in a deprived environment?...

Is an animal born with a blank slate or are all its behavior patterns inherited so that its future activities are largely predetermined? Does a mouse "know" from birth that, if it is to avoid being eaten, it must hide or run away from a cat? Do the parents of a young swallow "teach" it to fly northwards in spring to its breeding grounds? Biologists have long been puzzled by the question of whether animal, and human, behavior is inherited, acquired, or perhaps, as most ethologists now believe, a mixture of both.

Each animal species possesses a repertoire of different behavior patterns which are typical of its members. Some of these are found only in that particular species. For example, the songs and courtship displays of birds tend to differ considerably from one species to the next for the simple reason that one of their functions is to indicate to prospective mates to which species a courting male belongs. Other behavior patterns, such as how an animal walks and the way it scratches its head, are also very similar throughout a species but these tend to be much the same in closely related species, too.

Originally, ethologists called such stereotyped and rather constant features of behavior "fixed action patterns" because they were so much the same in all the members of a species. Some early ethologists thought that this constancy arose because such behavior patterns were inherited, and they often referred to such acts as "instinctive" or "innate," imagining them to be a completely fixed characteristic of the species, like the number of toes it has or whether or not it has hair. But referring to behavior patterns as inherited has led to a great deal of argument over the years: the so-called "nature/nurture" or "learning/instinct" controversy. Just how reasonable is it to split behavior up into learned and innate components? Is it useful to refer to behavior patterns as being inherited? Today, very few ethologists think that it is.

The fact that a behavior pattern is stereotyped in form does not necessarily mean that learning plays no role in its development. Copying from other animals is a very precise way of ensuring that actions are identical between them. The learning of language by humans illustrates this point well. The words on this page can be understood by almost any reader who knows English just because this learning is so precise. Tiny differences in the shapes of letters can make the world of difference to their meaning. Another good example comes from the way that young gulls peck at the bills of their parents which stimulates them to regurgitate the food they have brought back for them. In the Laughing gull, the beak is completely red; in the Herring gull it is yellow with a red spot on it and the pecking of the chicks is directed at this spot. The pecking starts early in life and chicks a few days old will only peck at models of heads and beaks that look like those of their own species. They also direct their pecks very specifically at the red parts of the

▲ **Pecking in Laughing gull** (*Larus atricilla*) chicks; (1) The chick aims a "begging peck" at the parent's beak, which it grasps (2) and pulls downward. The parent then (3) regurgitates food which the chick eats with a "feeding peck" (4).

▼ **Directing its peck** at the red spot, a Great black-backed gull (*Larus marinus*) chick begs for food.

▶ **Pecking patterns** in Laughing gull chicks. Day-old chicks peck at models of a variety of objects, but prefer red objects and those that are bill shaped. Older more experienced chicks show a preference for models of Laughing gull heads, but also peck considerably at the Herring gull head, the chicks responding to the red spot on the model.

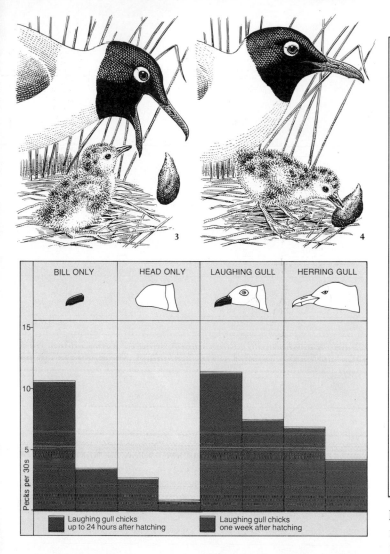

BILL ONLY	HEAD ONLY	LAUGHING GULL	HERRING GULL

Pecks per 30s

15-
10-
5-

Laughing gull chicks up to 24 hours after hatching

Laughing gull chicks one week after hatching

model. Does this mean that their response and the stimulus to which it is directed are built-in features of gull chicks? Careful study has shown that this is far from being the case and the behavior of the chick changes in quite a complicated way in the first days of its life. At first, they will peck at a great variety of objects, though red is preferred, as are rods that move to and fro like the adult's bill. It is only after some days of experience with their parents that they start to peck more at models of their own species than at those of other gulls. In other words, they have a tendency to peck at roughly the right things to begin with but it takes time for them to learn exactly what their parents look like and so direct their pecking only at models that resemble them closely. By then their behavior is certainly very fixed, but this is because of the way their predispositions have interacted with their experience: the result is not all due to learning nor is it all due to inheritance. Nature and nurture both have a part to play.

As behavior has been studied in greater detail, it has become apparent that the pecking of gull chicks is by no means exceptional. No matter what behavior pattern is examined, both heredity and environment have a part to play in its development. Learning is only one aspect of the way in which an animal's environment affects how its behavior develops, but it is an important one, and many very fixed features of behavior have been found to depend upon learning for their development. On the other hand, inheritance has a huge impact on the way that animals behave. We would not expect a snake to fly or an elephant to swim around underwater because their size and shape make these feats impossible for them. But the effects of inheritance are also much more subtle than this. All the behavior of animals is affected in one way or another by the genes that they possess. Many studies have been carried out to examine the differences in behavior between strains of animals and, in mice, for example, these have shown some to be placid and docile, others jumpy and nervous. These are all mice, the only difference between them being in the particular combination of genes that they possess, each strain having a rather different assortment of genes from every other one.

At its most subtle level, the difference between two animals may only lie in a single gene, just one unit out of the huge number that determine their inheritance, so that two groups of animals may be exactly the same except for this single point. Yet this one gene may affect many aspects of behavior. Fruit flies, possessing a gene called "yellow," are more sluggish in their courtship. Similarly, many of the genes which affect the coat colors of mice also affect their activity in one way or another. Indeed, it appears that nearly half the genes that a mouse has influence how active it is. But is this surprising? The activity of a mouse is, of course, bound to be affected by the length of its legs, the size of its muscles, how heavy it is, how clearly it can see and hear, how well fed it is, and many

The Inheritance of Learning

At a time when it was usual for those who studied behavior to think of different features as either learned or inherited, some experiments on the learning ability of rats were particularly striking. If hungry rats are put in a maze where there is food at the far end they will learn, over a series of tests, to take the correct turns to reach the food as soon as possible. But some of them learn much more quickly than others. What are the young of the next generation like if you mate those that learned fastest with each other and similarly those that learned most slowly? The young of the fast ones are faster than average and the young of the slow ones slower. By selective breeding for several generations, a fast line of "maze-bright" rats and a slow line of "maze-dull" rats can be produced. The obvious conclusion is that inheritance has an important role in the capacity to learn.

other factors. So, the many genes which affect these things will also have some impact on its activity.

Another way in which we can see whether inheritance affects behavior is by selective breeding. Any group of animals varies in numerous different ways, in behavior as well as in other respects. We can choose a particular characteristic, such as leg length, courtship speed or a liking for going through holes, and breed for it. In each generation, the best animals are mated with one another and then the offspring are studied to see whether there is an improvement from one generation to the next. Do legs get longer, or does courtship get faster or do the animals become fonder and fonder of holes? This will only happen if some aspect of the feature for which we are selecting depends on inheritance, otherwise the selection has nothing to work on. But, with behavior, the answer is clear: no matter what ethologists have chosen to select for, this characteristic has improved from generation to generation as a result of selection. In other words, all aspects of behavior—even those in which learning plays a very obvious role—are also dependent on the assortment of genes that the animal has.

Clearly, then, nature and nurture both have a crucial influence on all behavior. Even if learning is essential for the development of a particular feature, the inheritance of the animal is likely to affect its learning ability, and so its skill at achieving the result. At the other extreme, some behavior patterns may appear without any learning at all, yet it is a mistake to say they are simply inherited. All sorts of factors, such as how well nourished the animal is or whether it has seen a predator, may influence its likelihood of showing the action concerned. Its genes may predispose it to behave in a particular way, but the environment in which those genes develop may still be crucial in determining whether the animal really does show that action or not. It is as a result of ideas such as these that ethologists have ceased to ask questions about whether behavior is inherited or learned and, instead, started to examine the way that an animal's genes and the environment interact to develop the behavior that it shows.   PJBS

▶ **Many aspects of behavior,** such as the calls and holds used by these male Strawberry poison dart frogs (*Dendrobates pumilio*), which are fighting a duel over territorial rights, develop similarly in all members of a species. This does not mean that they are "innate." Both similarity of genes and the environment in which the young develop have a crucial role to play.

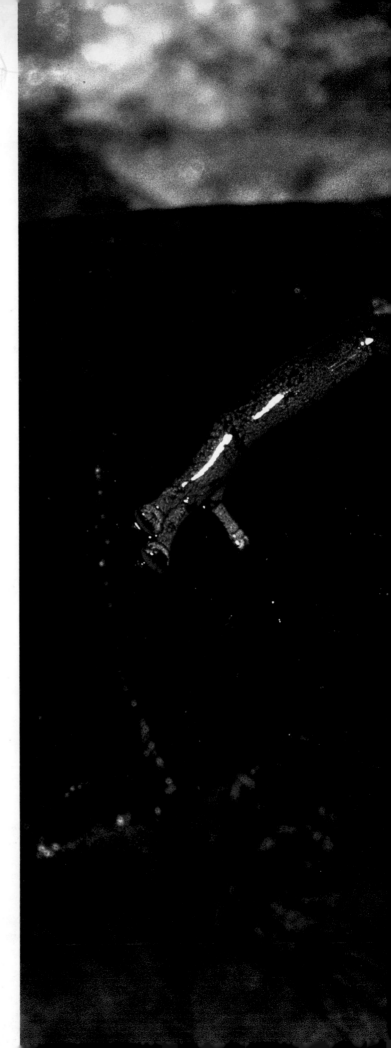

EVOLUTION OF BEHAVIOR

Inheritance of variation in behavior—experiments with fruit flies . . . Darwin and artificial selection. . . Selection of breeding sites in kittiwakes. . . Where do displays come from ?. . . Courtship in the peacock. . . Creation of new species. . . Learning to know your parents—fostering gull chicks . . . Behavioral and evolutionary relationships in birds. . . Evolution of courtship in the Great-crested grebe. . .

THE way in which an animal behaves is every bit as important for its survival as its color and general body form. Natural selection has shaped the behavior of different animal species in accordance with the demands of their different life styles; as the way of life of a species and its form change over evolutionary time, so does its behavior.

If the behavior of animals of one species is examined, it is often found that individuals differ—some are more active than others, some can run faster, some sing more complicated songs, some get into more fights. Often this variation in behavior turns out to have an inherited basis. This can be demonstrated with a selection experiment. For example, with a population of the geneticists' favorite animal, the fruit fly, the tendency of individuals to approach or avoid gravity is measured. If, in a series of generations, parents are selected from those individuals that avoid gravity, then a line of gravity avoiders is produced, while, if gravity approachers are selected the opposite result occurs. This demonstrates that some of the variation in the original population must have had an hereditary basis because otherwise it should make no difference which flies we select as parents.

Charles Darwin was well aware of the results of such artificial selection performed with animals put to use by humans—for example, in the enormous differences in behavior between different breeds of dog, all of which are members of one species, because of human selection for retrieving and pointing at game, herding sheep, hunting foxes, fighting other dogs and so on. The results of such artificial selection were important in forming Darwin's idea of natural selection.

Darwin suggested that, under natural conditions, some of the variability between individuals would affect their ability to survive and reproduce successfully. Those variants that were more successful would leave more surviving offspring and, hence, their frequency would increase in the population over successive generations. This is the main reason that behavior evolves and that different species have come to behave differently; because animals only breed with their own species, they do not exchange genetic variants with other ones and they

▶ **Walking on water**—mating run of the Western grebe (*Aechmophorus occidentalis*). As with most bird species, courtship involves a series of ritualized displays, some of which are similar in different related species.

▼ **Ritualized courtship dances** in the Great-crested grebe (*Podiceps cristatus*). (**1**) Head-shaking ceremony. (**2**) Dive and cat display. (**3**) Mutual greeting and "cat display." (**4**) Penguin dance.

Evolution of Courtship in the Great-crested Grebe

The displays of Great-crested grebes were studied by Julian Huxley in a pioneering project carried out early in this century. He found this species to have some of the most elaborate courtship and pair-bonding displays to be found anywhere in the animal kingdom. The courtship ceremonies consist of a series of behaviors which are performed with maximum intensity when a mated pair of birds reunites. Each behavior consists of a combination of postures and movements which appear to have evolved from simpler, less stereotyped patterns. Sometimes, the birds go through a head-shaking

can evolve characters independently in accordance with the demands of their surroundings.

This type of adaptive evolution is well illustrated by the kittiwake, which is unusual among gulls in that it nests on small ledges on precipitous sea cliffs. At these sites, the nests are safe from most predators, and this was probably the major reason for the evolution of cliff-nesting kittiwakes. The kittiwake shows a number of behavioral differences from ground-nesting species. When kittiwakes fight, they dart the head horizontally and try to grasp the bill of the opponent. Then they twist the head from side to side until the opponent is unbalanced and thrown off the nest ledge. The ground-nesting species usually try either to get above their opponent and peck down, or to pull the opponent. The lack of space on the ledges would make both methods of attack awkward for the kittiwake and, anyway, its twisting technique is an effective method of evicting another bird from a small ledge. This fighting technique means that the bill of the kittiwake has become an important stimulus in aggressive encounters and, consequently, kittiwakes, unlike other gulls, hide their bills when frightened. Young kittiwakes attacked by adults also turn to display a black band on the nape of the neck which is not found in the young of any of the

ceremony (1) in which the conspicuous crest is raised, and the pair face each other and shake their heads emphatically from side to side. This display has apparently evolved from the turning-away movements shown by a bird when it switches from aggression to appeasement. Sometimes, when a pair comes together, one bird dives while the other waits in the extraordinary cat attitude (2). The diver then emerges with its back to the other bird and settles close to it. Alternatively, the cat attitude may be part of a mutual greeting in which one bird shows the display while the other rises out of the water in front of it (3). There often follows the bizarre "penguin

dance." Both birds dive and reappear with bunches of weed in their bills (4). They swim towards one another and then spring upright and move together shaking their heads. This display probably evolved from behavior usually associated with nest building.

While it is not easy to determine the role played by each of these displays, they probably play a part in maintaining the bond between the birds. Studying them led Huxley to propose the idea that displays evolved from other behavior patterns through a process he called "ritualization," which involved them becoming more exaggerated and stereotyped so that they were easily recognized.

ground-nesting gulls. The kittiwake chicks must stay on their ledges until they can fly, so they are physically prevented from moving away from an attacking adult. The black nape seems to have evolved as an appeasement signal. Because the young of ground-nesting species can move around relatively freely, the parents learn to recognize their own young within a few days of hatching. Kittiwakes, on the other hand, do not recognize their own chicks, presumably because there is no danger that they will wander.

Displays used in courtship and aggressive encounters often seem bizarre to the human observer. Comparing the displays of related species, however, can often give us a clue about their evolutionary origins.

Behavior and Evolutionary Relationships

The fact that behavior evolves can be used to help disentangle the evolutionary relationship between different species. The evolutionary idea of relatedness is basically the same as human relatedness; it is a reflection of how recently two species have had a "common ancestor." It is possible to work out evolutionary relatedness by looking at the fossil record. Often, however, the fossil record is so poor that this is impossible and then relatedness can only be determined by a comparison between the characteristics of different species alive today; the more recently two species have had a common ancestor, the more characters they should have in common, because they have had less time to evolve characters independently. Therefore it is possible to place

species in a family tree according to their degree of similarity. Behavior patterns are just as characteristic of species as body form, and behavior can be a valuable guide to relationships.

This method has been used to help classify a group of fish-eating birds, the Pelicaniformes, which includes the pelicans and cormorants. Nine behavior patterns which have been studied are either present or absent in each species, and the number of patterns held in common by the different species can be used to construct a family tree (see below). The tree is arranged so that species with many behaviors in common, such as anhingas and cormorants, are closer in the family tree than species with only a few, such as pelicans and tropic birds.

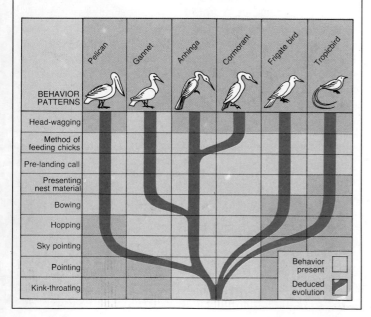

BEHAVIOR PATTERNS	Pelican	Gannet	Anhinga	Cormorant	Frigate bird	Tropicbird
Head-wagging						
Method of feeding chicks						
Pre-landing call						
Presenting nest material						
Bowing						
Hopping						
Sky pointing						
Pointing						
Kink-throating						

Behavior present
Deduced evolution

One of the most spectacular displays is the courtship of the male peacock. The courting male bows his head in front of the female and, at the same time, raises, spreads and vibrates his enormous ornamented tail. How could such an extraordinary display have originated? A strong hint comes from examining the peacock's close relatives. A female domestic fowl helps her chicks to find food by scratching the ground with her foot and then standing back and pecking at the ground, thereby showing the chicks where to peck. At the same time the hen produces a characteristic call. Cockerels, the males of the species, behave similarly when courting females. They scratch, peck and call in a similar way but without food being there. The female comes to look for food, and the male then bows and raises his tail in a sexual display. Many other pheasant species show a similar pattern of behavior and, in some species, the tail is large and decorated. In the mature male peacock, the original scratching and pecking movements have disappeared, and only the bowing and tail spreading components are present. Interestingly, young male peacocks do show the scratching and pecking behavior. So, in this group of birds, a behavior originally shown by females of one species when helping chicks to feed has been progressively modified by selection into the male display behavior in other species. The process by which features of behavior change during the course of evolution to become stylized and stereotyped is referred to as ritualization.

During evolution, new species form by the splitting of old ones. The isolated populations then evolve independently in their different environments, and their characteristics often diverge. If the geographical barrier subsequently disappears and the populations come back into contact, interbreeding may once again be possible. Matings between such distinct populations may produce inviable or sterile hybrids, however, and hence result in few surviving young. This is tantamount to saying that natural selection will favor individuals that do not breed with a member of the other population because these will leave more surviving offspring. In such circumstances, selection favors distinct coloration and displays in the two groups, so that they can readily recognize members of only their own population and mate only with these. Once this has happened, speciation has occurred. Sometimes, the differences in the displays are fairly subtle. For example, male *Sceloporus* lizards bob their heads when attracting females, and each of the seven species has a slightly different pattern of bobbing that attracts only females of the same species.

Sometimes recognition of members of the same species is based on learning by the young of the appearance of their parents. This occurs in two common species of gull, the Lesser black-backed and Herring gulls. Parents of these two species will accept eggs or chicks of the other species in the nest and then rear them as if they were their own. Such experiments show that, whereas chicks reared by foster parents of their own species invariably pick a mate of the same species, chicks reared by foster parents of the other species often form hybrid pairs with a member of the other species. So, in this case, the chicks must learn something about the appearance or behavior of the birds that reared them, and use this information when at sexual maturity they choose a mate of their own with which to breed.

LPa

▲▶ **The elaborate courtship display** of the male Common peafowl (*Pavus cristatus*)—the peacock—has probably evolved from a display originally used to call other peacocks to food. (1) In the male domestic fowl (*Gallus gallus*) and (2) the Ring-necked pheasant (*Phasianus colchicus*) food is often present during courtship and the male scratches the ground, pecks and calls. In all other pheasants, for example the Himalayan monal (*Lophophorus impejanus*) (3) and Burmese peacock-pheasant (*Polyplectron bicalcaratum*) (4) the tail is extended into a fan during courtship, and food is in general absent, but ground pecking still occurs. In the male peacock (5) the tail has become enormously enlarged and elaborately ornamented.

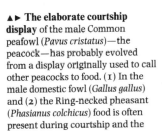

▲▶ **Fighting in gulls** has evolved differently depending on the type of nest site used. In kittiwakes (*Rissa tridactyla*), which nest on ledges ABOVE, fighting often involves twisting movements, whereas in ground-nesting gulls, such as Lesser black-backed gulls (*Larus fuscus*) RIGHT, opponents try to get above each other.

SOCIOBIOLOGY

Feigning injury to protect young. . . Altruism and its benefits to relatives. . . Behavior of the honeybee. . . Kin selection in scrub jays. . . Hunting and foraging in packs. . . Returning favors and cheating. . . The uses of alarm calling in birds. . . Avoiding breeding with relatives. . .

DURING the breeding season, the eggs and newly hatched young of many species of birds are taken by predators. To minimize the threat to their offspring, parents will often perform an elaborate series of behaviors to distract the attention of the potential predator. An adult Ringed plover, for example, will feign injury by pretending to have a broken wing. The bird uses the pretence to draw the enemy away from the nest or young. Such altruistic acts serve to increase the chances of survival of the offspring but often at some risk to the parent. If, however, the behavior results in an overall increase in reproductive success, that is, the distraction display achieves its objective, then any genes which increase the tendency to feign injury will be passed on to the next generation. Eventually, the trait will spread through the population at the expense of those individuals which lose their young because they failed to distract the predator.

Such acts of parental "altruism" are widespread throughout the animal kingdom. They illustrate how natural selection favors the parent which maximizes the contribution of its genes to future generations. Not all forms of altruism are so straightforward. When a worker bee encounters an intruder in its hive, it will sting the foreign bee to death. This is suicidal defense because the barbed sting is ripped from the worker's body and it dies as a result. Why should the worker commit the ultimate sacrifice when, as a sterile female, she has no offspring to protect, and why, for that matter, is she sterile?

For many years biologists interpreted the altruism of the worker bee in terms of the overall benefit to the population. It was assumed that individuals sacrificed their own success to increase the survival chances of the species. The mechanism suggested for the evolution of such behavior was termed "group selection." Recent theory and more detailed observations of animal societies, however, suggest that individuals do not behave for the good of the species but rather in their own selfish interest. How then does a gene for altruism become established in a worker bee?

It was in the early 1960s that William Hamilton developed his genetical theory for the evolution of behavior. His theory has revolutionized the study of animal societies, especially our understanding of social behavior. This approach is now often referred to as "sociobiology." He realized that, if selection favored the evolution of altruistic acts between parents and

offspring, then similar behavior might occur between other close relatives possessing the same altruistic genes which were identical by descent. In other words, individuals may behave kindly not only to their own immediate offspring but to others such as siblings, grandchildren and cousins. The more likely it is that the recipient contains the same genes, the more the donor would benefit from being altruistic, so that animals should help close kin more often than distant relatives.

Hamilton's theory makes it possible to understand the behavior and sterility of the worker bee. Whereas a fertilized human egg, whether male or female, has two sets of chromosomes, one from the father and one from the mother, the social insects of the order Hymenoptera, such as bees, have an unusual sex difference in chromosomal inheritance. Males develop from unfertilized eggs; they only have a single set of chromosomes. Females develop from fertilized eggs and so have the full complement of chromosomes. In humans and most other animals, a daughter will share half her genes with each parent as well as, on average, half with each brother and sister. But, in the social bees, a female will have more genes in common with her sisters than her own daughters. This is because sisters share all the genes they receive from their father (as he has only the single set of chromosomes) plus, on average, half the genes from their mother (with the normal two sets). Therefore, between sister bees, 75 percent of their genes are common by descent compared to only 50 percent between daughter and mother. The result of this genetic asymmetry is that a daughter will normally pass on more of her genes by helping to rear her reproductive sisters (future queens) than by attempting to rear her own young. Thus, the curious makeup of bees will favor the evolution of non-reproductive workers and hence their sterility. In addition, their suicidal behavior in protecting the hive can be seen as a means of ensuring the perpetuation of their genes through the future survival and reproduction of the sisters they have been helping to rear. The process through which this altruism has evolved is known as "kin selection."

▲ **Feigning injury,** a Ringed plover (*Charadrius hiaticula*) will lead a predator away from its nest at risk to itself.

▼ **Dinner time for lion cubs.** The lioness is allowing not only her own cubs to feed, but one from another litter in the pride.

▶ **Genetic relatedness and breeding advantages.** (**1**) In normal (diploid) species, such as mammals, offspring of generation 1 have half their genes in common with their mother and similarly, on average, half the genes in common with each other, having derived half each from common mother and father. (**2**) In social insects the mother has a normal (diploid) number of chromosomes, but the father half (haploid). Offspring (only daughters) have half their genes in common with their mother, but because they derive exactly the same genes from their father, sisters have three-quarters of their genes in common with each other. Thus in social insects, daughters are more closely related to each other than to their mothers. If a daughter social insect was to breed, then she would again only be half related to her own daughters. Therefore there is more benefit to their genes if daughters help raise further sisters rather than their own daughters. The genetic make-up of social insects thus predisposes them to evolve a social system in which sterile females care for siblings. In mammals, mothers, daughters and siblings are, on average, always 50 percent related, so it is immaterial for their genes whether they help raise siblings or produce their own young.

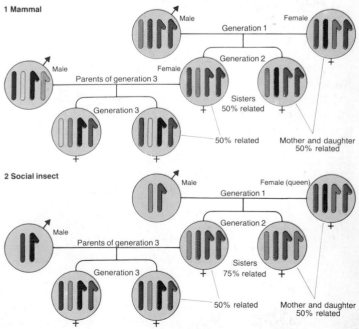

1 Mammal

Male · Female
Generation 1

Male · Female
Parents of generation 3 · Generation 2

Generation 3

Sisters 50% related

50% related

Mother and daughter 50% related

2 Social insect

Male · Female (queen)
Generation 1

Male · Female
Parents of generation 3 · Generation 2

Generation 3

Sisters 75% related

50% related

Mother and daughter 50% related

This new insight into behavior and the genetical implications of altruism stimulated an upsurge of interest in social behavior during the 1970s. Altruism and cooperation between animals came to be viewed in a new light as biologists realized the potential for kin selection. Detailed studies have now shown that the social organizations of many species of animals, particularly the birds and mammals, are based on kinship. For example, the Florida scrub jay is a communally breeding bird in which the offspring assist their parents in rearing subsequent broods. These are their own siblings and, by helping to rear them, they are in effect spreading copies of their own genes into the future. Similarly, among many species of primates and social carnivores, such as jackals, coyotes, lions and Brown hyenas, cooperation between individuals has been frequently linked to animals, particularly females, staying among relatives in the groups where they were born.

Kinship, however, is not essential for the evolution of altruism or cooperation. When birds forage in flocks or mammals hunt in packs, each individual may obtain more food than when feeding alone. The advantage is to the individual so that the behavior does not require that the other members of the group are related to it. Cooperative behavior of this type has an immediate payoff to each animal. In some cases, however, the cooperation can be more sophisticated. In the Olive baboon, a male will often solicit the help of an unrelated one in the troop to distract the attention of a third male that is guarding a receptive female. The soliciting male then takes over the female while the other two are engaged in a dispute. But why should one male help another to obtain a female? It appears that, in the future, the favor is returned. This type of altruistic behavior is known as "reciprocal altruism"—"you scratch my back and I'll scratch yours."

A potential difficulty with reciprocal altruism is that, once the first favor has been given, why does not the recipient cheat and not return it. One possibility is that, in animal societies where individuals can recognize one another, cooperation will only occur between those that are known to reciprocate. Cheaters will be ostracized and ultimately lose out.

One of the many criticisms of the sociobiological approach is that it is all too easy to come up with adaptive explanations for behavior. Armchair theorizing can result in a multitude of hypotheses but no data with which to test them. This problem is illustrated by the arguments surrounding the function of alarm calling in birds.

In many small birds individuals will give an alarm call when an aerial predator, such as a sparrowhawk, is hunting in their area. Different species, such as the blackbird, Great tit and chaffinch, give a similar call. The note covers a narrow frequency range and is high pitched. To human ears, it sounds like a thin

▲ **No escape**—a pack kill by African wild dogs (*Lycaon pictus*). After a cooperative hunt, three dogs share in the killing—one holds the lips, another gets the prey by the throat and the third disembowels it.

▶ **Home in the prairies.** Black-tailed prairie dogs (*Cynomys ludovicianus*) live in townships called "coteries"—see box.

▼ **Watch out there's a hawk about.** A blackbird gives an alarm call from cover when a bird of prey is approaching and this warns all other birds in the vicinity.

Avoiding Inbreeding

Many species of birds and mammals have a complex social organization based on kinship. Breeding groups frequently consist of related individuals. One of the dangers of having relatives of the opposite sex in close proximity is that it can increase the possibility of inbreeding. Such matings may be harmful, in an evolutionary sense, because the offspring produced from an incestuous union are often weak and defective and suffer from inbreeding depression. In other words, an animal's reproductive success would be higher if it mated with an unrelated individual. Field studies of a number of species have shown that behavior patterns can evolve which enable animals to retain the advantages of a social organization based around kin while minimizing the possibility of inbreeding. One of these species is the Black-tailed prairie dog.

Black-tails are large, day-active rodents which live in the northwestern part of the United States. They live in large colonies consisting of adjacent family groups known as "coteries." Each coterie usually contains one breeding male, three or four breeding females, young of the year and a number of yearlings. Family members are friendly towards one another but normally hostile to neighbors.

Both sexes usually start to breed in their second year. But, despite this close family structure, prairie dogs rarely mate with close genetic relatives. A number of behaviors occurs which may have evolved to minimize the risk of inbreeding. First, while the females born into a group remain there, the young males disperse before breeding. Second, a breeding male will usually leave his group before any of his daughters are mature and sexually receptive. Third, when the father is still present, his young daughter is less likely to come into heat than if he had departed. And finally, a receptive female will actively try to avoid mating with any close relative still present in the coterie.

whistle and its source is often difficult to locate; it has a ventriloquial quality.

What is the function of this call? Over the last 20 years, a large number of explanations has been put forward. Some believe that the call has evolved through kin selection and serves to warn close relatives of danger. Others consider that the behavior has evolved through direct advantage to the individual but, even so, there is a variety of different hypotheses. Does the caller take a risk and assist unrelated neighbors because such behavior will be reciprocated in the future? Alternatively, is the caller saving time and energy by announcing to the predator that it has been spotted and that it is not worth pursuing a vigilant bird? Does the ventriloquial nature of the call actually divert the predator away from the caller? It has even been suggested that alarm callers are selfishly manipulating other members of the population by warning them of danger but not its location. The caller minimizes its own risk by causing panic among the others.

Many of these hypotheses are plausible ones but, to date, we do not know the function of alarm calling in birds nor whether its effect differs from species to species.

One of the objectives of sociobiology is to incorporate the study of human behavior within the same framework. This has generated a great deal of controversy. One of the tenets of sociobiology is that many of the individual, sexual and racial variations in human behavior are genetically determined and adaptive. Many biologists question this viewpoint and stress the importance of social and cultural forces. Human social behavior is extremely flexible and reflects more the importance of cultural than genetic transmission. Nevertheless, the controversy lingers on.   PG

Social Organizations

MANY animal species live in groups: the hives of bees, the schools of fishes, the herds of deer, the flocks of birds. Some of these exist simply because food is clumped and all must visit the same area to feed, or because the animal surrounded by others gains protection from predators. These were two important reasons why groups first arose through evolution. But most groups are not amorphous masses of independent individuals. Especially among the social insects and the mammals, different animals may have different roles to play within the group according to such factors as their caste, their age, their sex or their relationships with others. The principles of sociobiology help us to understand much that earlier biologists found perplexing about these societies. Why are some insects sterile? What are the advantages of the dominance hierarchies so often found in animal groups? Why do groups differ so much in their size, the distances over which they travel and their sex ratio? These are a few of the fundamental questions about social groups to which recent research has brought answers.

One important feature here is the way in which young animals develop within the group from infancy, through adolescence to adulthood, their relationships with others changing as they mature. Learning plays a key role here, as it does in another feature of many social animals, that they copy or imitate one another so that useful aspects of behavior are passed between them culturally. These elements of social life are singled out for treatment on their own as they say so much about what being social means to an animal. Relationships which involve more than one species are also described, partly because they involve some fascinating natural history but partly, too, because they give us an insight into how unlikely arrangements can arise. At first sight, this is perplexing because animals behaving in their own best interest would hardly be expected to assist those of a totally different species.

We end with a short article as an epilogue, musing on the rich variety of animal behavior and asking just how relevant understanding it is to that of our own species. The conclusion is the same as that which Darwin reached in his *The Descent of Man and Selection in Relation to Sex* (1871) though we now have many more examples with which to make the point: the differences between animals and humans are of degree rather than of kind. If we study the behavior of animals, we find aspects of our own mirrored there in simpler form. Looking at the reflection may help us just a little to understand ourselves better but, even without that, it is an image of great beauty and is well worth studying for that alone.

◄ **Living together**—a hermit crab (*Eupagurus* species) covered in sea anemones (*Calliactis parasitica*).

INSECT SOCIETIES

*The caste system in termites... Queens and workers in ants,
honeybees and stingless bees... Changing duties in honeybee
workers... The control of colonies by queens... Bee
pheromones... Swarming in bees... Recognizing brood...
Transferring food within a colony... Defending nests from
predators... Scent trails of termites, ants and bees...
Mating flights in bees... Alarm pheromones... Scent glands
of the worker bee...*

W<small>E</small> take it for granted that humans live in groups which
are thought of as societies. In other words, we voluntarily
associate and cooperate with others of our kind for company,
for reproduction, and to achieve common ends born out of common needs and interests. And, it is not hard to see that many
species of other mammals may do the same according to their
requirements and the demands of their environment. It is less
easy to imagine that the lowly insects, with their simpler
nervous systems and apparently less complex modes of life,
might also maintain interacting societies, even though bee
keepers, for example, have made use of honeybee social
behavior for many centuries.

There are two main groups of social insects. The termites
belong to the order Isoptera, of which there are no solitary
representatives, and the social bees, wasps and ants belong to
the order Hymenoptera. Social life in the insect world is advantageous in providing for cooperative defense and an efficient
division of labor among the members of a colony. These individual members may be similar in body form or they may
belong to well-defined castes.

There are three castes in the termite colony: the reproductives (queen and king), workers and soldiers. Among the
termites, males as well as females belong to different castes but
only the queen and king, of which there is one pair per colony,
and the soldiers have completed their development, while all
the remaining individuals are immature forms which are
analogous to the worker caste of other social insects. The
soldier termites possess enlarged head capsules, either with
large mandibles or with a frontal projection through which
defensive secretions are emitted. After mating, the queen
changes dramatically in body shape and form, and her abdomen may increase to five or ten times its original size.

Among the bees and wasps, there are species representing
a wide range of increasing sociality. Caste production in ants,
bees and wasps is limited to the female sex only and the
immature stages perform no duties.

In some of the less-developed social insects, such as certain
wasps and bees, the queen, the fertile female, is distinctly larger
than the workers which are sterile females, but a wide variation
can occur in the size and physiological characteristics of each
caste. At certain stages of her life, the queen is able to undertake
all the activities normally performed by a female of a solitary
species.

In contrast, there are marked differences in form and function between queen and workers among the ants, honeybees
and stingless bees. An ant colony is comprised of one or more
fertile females and many workers, all of which are sterile

▶ **A buzzing hive**—honeybees and
brood cells containing larvae

▼ **Back from "shopping,"** a social
wasp (*Polistes testaceicolor*) passes a
bolus of food to another wasp,
which will feed it to the larvae.

females. Workers commonly vary in size. Ant queens usually develop wings, which are shed after copulation, but workers never do. To found her colony a queen's tasks include: cell building, egg laying, brood rearing and defense. Hence, she retains a wide behavioral repertoire until her workers are able to help; thereafter she concentrates on feeding and egg laying.

Queens of honeybees and stingless bees are unable to survive alone or to perform any duties except mating and egg laying; their colonies survive throughout the year and reproduce by swarming. A queen honeybee lacks the pollen baskets, long tongue and barbed sting of the worker. All the workers in a honeybee colony are in form identical, yet their ability to perform certain tasks, such as comb building and brood feeding, depends on the state of development of the relevant glands. In general, with increase in age, worker honeybees undertake four overlapping series of duties: cell cleaning; comb building and brood feeding; nectar reception, pollen packing, removing debris, guarding; and foraging. They exhibit great adaptability in the tasks they undertake, however, according to the needs of their colony.

The queens of social Hymenoptera influence the behavior and physiology of their workers and prevent them from becoming reproductives, either by physical domination or by secreting chemicals called pheromones.

In certain species of *Polistes* wasps, in which several mated females collaborate to found a colony, a dominance hierarchy headed by the wasp that becomes "queen" is soon established between them, the more dominant wasps receiving most food. They also eat any eggs produced by subordinates. This feeding advantage to the queen and dominant individuals is reflected in their greater ovary development. The lowest-ranking individuals in the hierarchy forage and build the cells of the nest.

In the most highly developed species of social insects, physical dominance by the queen gives place to more subtle dominance by pheromones. As long as worker honeybees can maintain contact with their queen, the rearing of additional queens is inhibited. But, in the absence of a queen, or when the supply of queen pheromone in a honeybee colony becomes inadequate because the queen is aging, her workers rear a new queen.

Worker honeybees can recognize the presence of their queen by the volatile pheromones she produces and, when the queen is stationary on the comb, many of the nearby workers face towards her. The composition of the so-called "court" of workers is constantly changing, however, and few remain with it for more than a minute or so. When in the court, the bees facing the queen sometimes lick her body and frequently palpate her with their antennae. While doing so, it seems probable that the surfaces of the antennae become coated with queen pheromone.

As a result of contacting the queen, a bee leaving her court is stimulated to greater activity and, for the next few minutes, ranges widely throughout the brood area of its colony. Such a worker is especially attractive to other workers which initiate antennae contact with it, presumably because they can detect the presence of queen pheromone on its antennae. Workers contacted by those that have just left the court, in turn palpate the antennae of others. The frequency of mutual palpating with the antennae, together with the continual movement of bees from the court, ensure that queen pheromone is adequately dispersed throughout the colony.

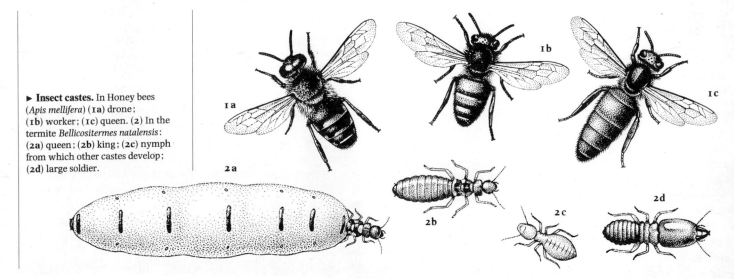

▶ **Insect castes.** In Honey bees (*Apis mellifera*) (**1a**) drone; (**1b**) worker; (**1c**) queen. (**2**) In the termite *Bellicositermes natalensis*: (**2a**) queen; (**2b**) king; (**2c**) nymph from which other castes develop; (**2d**) large soldier.

Pheromone from immature queens is similarly distributed. Immediately after a bee has visited cells containing immature queens, it grooms itself thoroughly which may help distribute any pheromone over its body. Soon afterwards it makes antennal contact with other workers. A worker bee also obtains pheromone from a queen larva by licking it; after such a visitation, the worker concerned often distributes the pheromone in food it donates to other bees.

Colonies also produce queens prior to swarming when the worker bees are not obtaining sufficient queen pheromone. This situation is resolved when a colony swarms; the old queen leaves with about half the bees to seek a new home, and the new queen heads the mother colony.

Queen pheromones are also partially responsible for inhibiting ovary development of honeybee workers. But, when a queen is removed from a colony, the worker ovarian activity continues to be inhibited by a pheromone produced by the brood, so worker egg laying does not run riot, and colony cohesion is maintained as long as the brood of the removed queen still remains. By the time this brood has all emerged, a new queen, which will soon produce a brood of her own, will probably have been reared.

Like honeybees, the queens of certain vespine wasps produce pheromones that maintain the stability of the colony and inhibit workers from laying eggs; if the queen is removed, the worker wasps become pugnacious and cease to care for the brood. In some ant species the queens' influence inhibits or

reduces queen rearing, again, probably by pheromones.

Termite caste production appears to be regulated by the number of individuals of the caste that are already present and the size of the colony. For example, the presence of one or more soldiers can influence the formation of additional ones, and the proportion of individuals that belong to the soldier caste may remain similar irrespective of colony size. The presence of a pair of reproductive termites inhibits the production of replacement reproductives in a colony, although this inhibition may be insufficient when the colony exceeds a critical size. Inhibition appears to be caused by a pheromone released from the anus of a primary reproductive. This pheromone is circulated in food; the larvae take in the pheromone through their mouths and, in turn, pass it on by anal release.

The ability of social insects to recognize their brood, its stage of development, sex and caste, is vital if the brood is to be correctly fed and cared for.

In a hornet colony, mechanical stimuli are important in

Nasonov Scent Gland of the Worker Honeybee

On the dorsal side of its abdomen, the worker honeybee possesses the Nasonov scent gland (named after the Russian who first described it in 1882) which it exposes (colored red BELOW) by flexing the tip of its abdomen. The pheromone produced attracts other bees to the site.

There are three main circumstances in which bees use this gland. Firstly, it is used to mark a source of water on which they are foraging so that it can be more readily discovered by bees they have recruited, especially when the water source does not have an odor of its own. Secondly, when a worker bee regains its hive or nest after being temporarily lost, it stands at the entrance, exposes its Nasonov gland and disperses the pheromone by

fanning its wings; this helps guide other bees home. Thirdly, Nasonov gland exposure, accompanied by fanning, is used extensively during the swarming process. The first workers from a newly issued swarm to settle on a support near their old home release Nasonov pheromone which attracts the queen and workers that are still airborne. Scout bees use Nasonov pheromone to mark the location of a potential nest site and many of the swarming bees expose their Nasonov glands when entering it. Possibly the pheromone may also be used to help maintain the cohesion of the swarm during flight.

The Nasonov pheromone is comprised of seven chemical components. Tests have shown, however, that only three of these are necessary to attract clustering bees and, indeed, that a mixture of these three in equal proportions is more effective than a mixture of the seven components in their naturally occurring proportions.

inducing brood care. The hornet larvae produce vibrations during their hunger movements and these induce workers to feed them. In addition, the members of a hornet brood produce a pheromone that stimulates workers to incubate them.

The brood of ants are grouped in chambers within the nest. The workers clean and feed them and transport them about the nest so they are maintained at favorable temperatures and humidities. Contact pheromones dispersed in the cuticle are of major importance in brood recognition by worker ants, and probably enable them to distinguish between broods of different castes and sex. In at least some ants, the texture and turgidity (internal pressure) of the brood skin are important means of recognition.

A pheromone is also the primary brood recognition signal in a honeybee colony but physical characteristics of the brood are of relatively little significance. Worker bees can even discriminate, through the wax capping of cells, between pupae of different ages. The cell type is also important in enabling rec-

ognition; differences in the size and odor of drone and worker cells enable bees to distinguish between them.

The transfer of food between individuals is of great importance in the organization and cohesion of the more highly developed insect societies; the food may provide a medium by which many of the pheromones are transferred among the colony members. In a honeybee colony, food transfer occurs only between adults but the young of wasps, ants and termites produce secretions attractive to adults which respond by feeding them. Worker vespine wasps sometimes squeeze their larvae in their mandibles to induce them to secrete drops of fluid, rich in sugar and proteins, from their salivary glands.

Forage brought back to the nest can be rapidly dispersed

among the members of a social insect colony and the transfer of food between bees may itself also work as a form of communication. In a honeybee colony, changes in incoming food supplies effect brood rearing, ripening and storing of honey, wax secretion and comb building. The concentration of the food being circulated determines the threshold of nectar sugar concentration that is acceptable to foragers.

A honeybee that is receiving food inserts its tongue between the mouthparts of the bee that is giving food which opens its mandibles and regurgitates food between them. During feeding, the antennae of both giver and receiver are constantly moving and palpating each other; this helps the bees to orient themselves to one another, and to communicate. The eagerness with which the food is transferred is reflected in the intensity with which the bees palpate each other.

Studies of food transfer between individual wasps have demonstrated that the receiving wasp needs to maintain correct palpations on the mandibles of the donor for it to continue regurgitating food. The aggressive behavior of dominant bees or wasps induces subordinates to regurgitate food to them. Possibly food transfer evolved from such behavior.

In common with many other animal reactions, the begging and offering responses are released by only a few out of a number of possible stimuli. Experiments with worker honeybees have shown that the presence of severed heads alone is enough to release either response. The scent of the head is an important stimulus, and heads elicit a greater response if taken from bees belonging to the same colony as the offering and begging bees. The antennae on the head provide an important contact stimulus in releasing both begging and offering. It is even possible to simulate the effect of antennae by inserting two pieces of wire, of approximately the same length and diameter as real antennae, into heads with no natural antennae.

The brood and stores of food in the nests of social insects offer tempting prizes to predators. Often, the most serious enemies against which social insects need defense are members of another colony of the same species. Members of the same

colony share a common odor which differs from that of other colonies and arises in part by adsorption on to their body surfaces of the odors of their nest and food stores. Would-be intruders are recognized by their behavior and by strange odors. For example, the characteristic swaying flight of would-be robber honeybees, enables the guard bees to recognize them but, at close quarters, the guard bees confirm the identity of intruders by scent.

The extreme importance of defense to social insects has led to many types of defensive behavior, and "alarm pheromones" are part of the defense armory of most, if not all, social insect colonies. They have three main functions: to alert the colony; to release aggression; and to mark the target to be attacked.

Those termites that seek food beyond the protection of the nest, tunnel or gallery they have constructed, employ scouts that lay a scent trail to the food with a glandular secretion. Even for those termites that seek food while protected by covered runways, "trail pheromones" appear to be important, because the stronger the trail, the wider and higher the tunnel walls that are built.

Ant foragers that discover a particularly rewarding source of food deposit a pheromone trail by daubing the tips of their abdomens into the ground or surface on which they are moving. This trail is followed by their nest mates which, in turn, reinforce the amount of pheromone present. As the food supply is depleted, fewer individuals use the trail and the odor diminishes.

Pheromone trails are also produced by some stingless bees to direct their sisters to a source of forage. When a forager has found a rewarding food supply, she stops every few meters on her homeward flight and deposits a secretion from her mandibular glands on a pebble, lump of soil or leaf, forming a trail that others can follow.

A successful honeybee forager indicates the location of a source of water or a bountiful food supply by the dances it performs on its return home, and does not form a trail, but it may

Alarm Pheromones

A honeybee that is alerted to danger reacts by elevating its abdomen and opening its sting chamber to release alarm pheromone which it disperses by rapidly fanning its wings. The alarm pheromone alerts other bees to assume aggressive postures, with mandibles agape and front legs raised, ready to fly to the attack at the slightest provocation. When attacking an intruder, a worker honeybee protrudes its barbed sting and thrusts it into the enemy. If the enemy has a soft skin, the barbs of the sting become embedded in it and, in attempting to withdraw it, the sting is torn from the attacking bee and left

in the intruder. For the next few minutes, the alarm pheromones continue to be released by the sting glands and so guide other defending bees to the target.

Some races of honeybee are noticeably more aggressive than others. The Africanized bee currently spreading in South and Central America responds to colony disturbance more quickly, in greater numbers and with more stinging than its European counterpart. Available evidence indicates that this is due to the Africanized bee's enhanced response to alarm pheromone, rather than to the production of greater amounts of it.

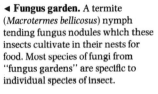

◄ **Fungus garden.** A termite (*Macrotermes bellicosus*) nymph tending fungus nodules which these insects cultivate in their nests for food. Most species of fungi from "fungus gardens" are specific to individual species of insect.

► **Bringing home the compost.** Leaf-cutting ants (*Atta* species) carrying cut leaves which will be used in the fungus garden as a basis for growing fungi.

▼ **The internal structure** of the nest of the Brown garden ant (*Lasius niger*) showing detail of the brood chamber.

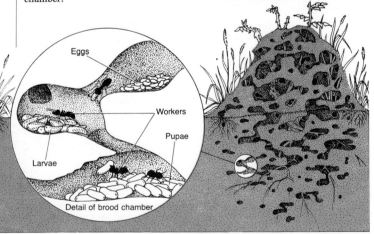

scent mark the location itself with a pheromone from the Nasonov gland.

In many termite species, the female "calls" her prospective mate. After a short flight from her maternal nest, she lands, and elevates her abdomen to expose her glands. One set of glands on the back, the tergal glands, releases pheromone that attracts males from up to 20cm (8in) distance, and the pheromone from the gland on the sternum, which is effective only up to 3cm (1.2in), maintains the couple in close proximity.

Bumblebees may use a specialized scent trail to facilitate mating. Males of many bumblebee species fly along a network of interwoven flight routes that link sites, such as a leaf, twig or base of a tree, near to which they have deposited a scent from glands on the upper mouthparts. The type of sites chosen for flight routes, and their height above ground, differ from bumblebee species to species. Even where the flight routes of different species cannot be separated in space, they can be distinguished by odor because the pheromones that males use to mark sites in the flight routes are specific to the species.

The wide dispersal of scented sites along flight routes provides large areas that are attractive to virgin bumblebee queens, so that the sexes are able to meet. When the queens are in the vicinity of the visiting sites, they elicit male sexual responses. The initial reaction to the queen is probably entirely visual, and the males react most to queens or models of queens of the color and markings of those of their own species. After the initial visual response, the correct odor is important in eliciting contact and attempted mating.

Males of some species of vespine wasps also appear to establish distinct flight paths. In hilly or mountainous regions, honeybees also have special mating sites known as "drone congregation areas." On entering a drone congregation area, a queen is quickly mated, often with several drones in succession, and is soon able to return to the safety of her nest.

JBF

VERTEBRATE SOCIETIES

IT is easy to take for granted the fact that many animals live
in social groups. A herd of cows, a school of fishes, a flock
of birds are commonplace, but why should so many kinds of
fishes, amphibians, reptiles, birds and mammals live in groups?
What kinds of groups do they live in? How do they manage
to stay together, particularly if they live in dense undergrowth
or murky water, where communication may be difficult? It is
only in recent years that some of the answers to these questions
have begun to emerge.

Some groups of animals are merely temporary gatherings.
Many birds flock during the winter, but the flock splits up at
the beginning of the breeding season. Also, in many groups,
the individuals are anonymous: the treatment that a bird in
a flock of starlings, or a fish in a shoal of sardines, receives from
its fellow group members does not depend upon its individual
identity. By contrast, in other kinds of animals, individuals live
their entire lives within just one group, and come to know the
idiosyncracies of every other group member. A female baboon
will stay from her birth until her death in the same group of
perhaps 60 animals. She will get to know the other females
in the group intimately—her mother, grandmother, aunts, sisters and so on. Her acquaintance with males in the group will
be more fleeting because males leave the group into which they
were born at adolescence and, from then on, come and go
between groups.

In many kinds of animals there is a close link between group
life and reproduction. This is not surprising: in most animals
the two sexes have to come together and cooperate to mate,
and the very acts of coming together and cooperating form the
essence of social life. Some animals, such as many frogs and
toads, meet only briefly to mate, lay fertilized eggs and then
depart, leaving the eggs to develop on their own as best they

**▲▶ Aspects of vertebrate
societies.** (1) A male ostrich
(*Struthio camelus*) incubates a large
number of eggs laid in his nest by a
number of females (seen behind),
which show little further interest in
their brood. (2) A family group of
Cotton-top tamarins (*Sanguinus
oedipus*), where older offspring delay
their own breeding by staying in the
family unit and caring for younger
siblings. An infant (2a) is
transferred from its mother (2b),
where it has been suckling, to an
older offspring helper (2c) who will
share the carrying duties; the twin
of the first infant is groomed by

another helper (2d) while it sits on
the back of its father (2e).
(3) Relationships in Savanna
baboons (*Papio cynocephalus*). Adult
male (3a) is the "special friend" of
female (3b), which enhances his
chances of mating when she is on
heat. Male (3c) is trying to become
integrated into the troup by
attempting to form a friendship with
(3b); he could dominate her on his
own, but not when she is supported
by the attendant male (3a).

"Friends"

Rhesus monkeys, baboons and
other monkeys that live in
large, complicated groups often
form alliances. Two adult male
baboons may help each other to
drive away another male from
an attractive adult female that
he is courting.

Alliances are also important
within families. An infant
enjoys the support of its
mother, aunts and other sisters,
together with any male
relatives that have not yet left
the group, if it should run into

a dispute. Infants from high-
ranking matriarchies
themselves become high
ranking precisely because the
other members of their
matriarchy are so good at
helping them in their disputes.
They come to the infants' aid
more readily and more
effectively than the members of
low-ranking matriarchies, and
more of them may offer this aid
in the first place. Also, if there
is a dispute between two
daughters within a maternal

lineage, the mother is likely to
come to the aid of the younger
rather than the older daughter.
This is why younger daughters
have a higher rank than their
older sisters.

Rhesus monkey and baboon
females also seem to have "best
friends" among the adult
males. A best friend is an adult
male that sticks by a mother,
particularly after she has given
birth, and by the infant as it
grows up. The male and mother
will tend to groom each other,

in preference to other members
of the opposite sex; the male
will occasionally play with or
carry the mother's latest infant,
and will prevent other animals
from taking the baby away
from its mother. It may be that
the male is the infant's father
but, because the animals are
promiscuous, this is uncertain.
It may also be that, by
associating with the female in
this way, the male is improving
his chances of mating with her
in the next mating season.

can. In many species, however, either one or both parents stay to look after their young and, when this happens, a social group results, even if it is only short lived. In mouth-brooding cichlid fishes, such as *Tilapia*, the mother carries the fertilized eggs and the newly hatched young in her mouth. Many species of birds live in families during the breeding season; both parents stay with the brood and feed them until they all disperse at the end of the breeding season. In some of these birds, such as swans, the same male and female pair each year; that is, they are monogamous throughout their breeding lives. In other birds, such as gulls, enormous numbers of families cluster together during the breeding season and so form a colony.

In species where the youngsters leave their parents early in life, the social group remains small and simple. Where they stay for longer, however, the groups become bigger and more complicated. The South American marmosets, which are small monkeys, live in extended families, each of which contains the mother, father and several offspring of different ages. The older ones help their parents to feed and carry their younger brothers and sisters whose chances of surviving may therefore be improved. This system also gives the older brothers and sisters valuable experience in parental care, and helps them to become better parents when they set up families of their own (see also Parental Behavior).

If the offspring remain in the group into which they are born, not for just a few breeding seasons, but for life, then large and elaborate groups result, containing many breeding animals. Among mammals, usually only the female offspring remain in the natal group, whereas the males leave. Some animals in which this happens, including many antelopes, deer and sea lions, live in harems, containing just one adult male, several females and their young. The extra males either live on their own or with other males in bachelor groups. In other mammals, the females again stay for life in the group into which they are born, but several males move in from other groups and live with them. Baboons, wallabies, lions, hyenas and Rhesus monkeys are all like this, and, as a result, their social life can become very complicated.

Life in a group has its advantages and its drawbacks. Birds, such as White-fronted geese, spend more of their time looking out for predators if they are in small flocks than if they are in big ones. The bigger the flock, the more time each goose can devote to valuable activities such as feeding. It is usually safer in the middle of a group than at the edges, because it is at the edges that predators, such as birds of prey, sharks and lions, most often capture their prey. If one of these predators appears, the threatened group may well become more compact, the individuals moving towards the center of the group and thereby escaping from the dangers at the edge. If animals live in groups, they may also be better able to deal with predators than on their own, simply by taking part in group defense. Birds will often cooperate and mob predators. The more birds there are in the mob, the more successful they are likely to be in driving away the predator.

An animal may do better at finding food when it is living in a group than it would if it were living on its own. Pelicans will often swim in formation along a lake when feeding, and dip their huge bills into the water in unison. A fish that swims

away from one pelican is likely to swim straight into the open beak of another. By hunting in packs, wolves and African wild dogs are able to capture animals many times their own size, such as moose or zebra, which would be far too big for them were they to hunt on their own. Animals may also benefit from feeding in groups even when their food is much smaller in size than themselves. It is rather difficult for a sparrow or a rabbit to disguise the fact that it is feeding, and nearby group members can quickly detect when one of their colleagues has discovered a new source of food, and so cash in on the other individual's fortunate discovery. An animal living on its own, has to find food unaided. On the other hand, it can eat the food, once food is found, unhindered and so the disadvantage may cancel out the advantage.

Group living may involve other disadvantages compared with solitary animals. If chimpanzees feed in large groups, each chimpanzee does worse than if it feeds in a small group. A large group of animals will probably be much more conspicuous to a predator than would a single animal. What is more, once a predator becomes attuned to one kind of prey, it may readily find other prey of the same species nearby, even if they have tried to hide.

When animals live together in the same group for any length

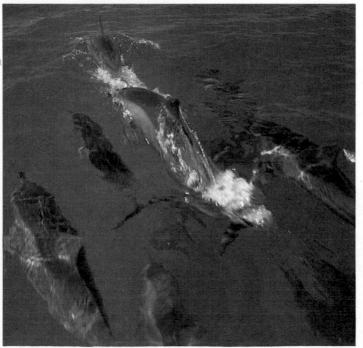

▲ **Group life on the African plains.**
Many species living in open habitats
have evolved social lifestyles.
Shown here are groups of ostrich
(*Struthio camelus*), gemsbok (*Oryx
gazella*) and Plains zebra (*Equus
burchelli*) visiting a water hole in
Etosha National Park.

▶ **Riding a bow wave,** a school of
Spinner dolphins (*Stenella
longirostris*) swim in front of a boat.
During the day, these animals live
in tight social groups of 10–100
individuals, only spreading out at
night to feed.

▼ **Pecking order in domestic fowl.**
(1) In small groups of birds, there is
what is known as a "peck-right"
hierarchy in which a low-ranking
bird never pecks a higher-ranking
one. Bird A will peck B, C and D,
bird B will peck C and D and bird
C pecks D. Such a dominance
hierarchy enables the dominant
individuals to have better access to
resources such as food and roosting
sites.
(2) In larger groups of 10 or more
individuals this simple linear
hierarchy breaks down, with some
lower-ranking birds pecking one or
two higher-ranking ones. Usually
these individuals are part of
alliances of related animals
which *en masse* are able to
intimidate higher-ranking
individuals.

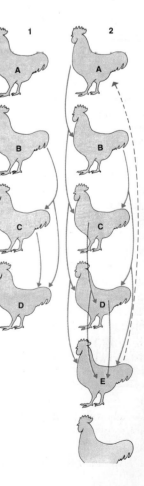

of time, the relationships between them can become subtle and complicated. Animals come to recognize other members of their group as individuals, and to tailor their own behavior to the individual peculiarities of the others. Animals from the same group may often compete with one another for something that is especially desirable: chickens for grains of wheat, or baboons for a ledge on a cliff where the group sleeps at night. Disputes are not always resolved by free-for-all fights. Sometimes one animal may simply defer to another, and the dispute is resolved without recourse to fighting. If a farmer puts several chickens together for the first time, they will, at the outset, peck any other chicken with which they come into conflict. Soon, however, the chickens learn to recognize one another, and each bird finds out which individual it can peck without retaliation, and which it would do better to avoid. Soon, a "pecking order" appears. The bird at the top pecks any of the others with which it has a dispute, and they in turn avoid the top-ranking bird if ever they face a potential conflict with it. The second-ranking bird will peck all but the top-ranking bird, which it will avoid, and so on down to the bottom-ranking bird which will avoid all the others.

Pecking orders are not just restricted to birds. They are particularly interesting in mammals but, because mammals do not literally peck each other, the mammalian equivalent of pecking order is called a "dominance hierarchy." Dominance hierarchies are rarely as simple as the straight-line pecking order of chickens. Some of the most elaborate hierarchies occur in Rhesus monkeys and baboons. It is quite common to find that one animal, say, a nearly mature male, is higher ranking than another, say, a slightly younger male, which in turn is higher ranking than a third, say an adult female. The adult female may herself, however, be higher ranking than the nearly mature male, so that there is a loop in the hierarchy. It could well be that this adult female is the nearly mature male's sister, and that he defers to her, whereas he does not defer to individuals to which he is not related.

In fact, family ties are very powerful within groups of Rhesus monkeys and baboons, and it is the combined effects of family loyalties and dominance hierarchies that makes Rhesus monkey society so intricate. Females spend all their lives in the group into which they are born, so that a Rhesus group comes to contain several female lineages. The members of a maternal lineage tend to stay clustered together, so it may be possible to see, sitting next door to one another in a group of monkeys, the matriarch, her daughters, her granddaughters, and even her great-granddaughters. Some lineages are powerful, others weak. This means that all the members of one lineage are higher ranking than all the members of another. Within a lineage, each mother is higher ranking than any of her daughters, but all except the very young daughters of a mother are ranked in the opposite order to their age: the oldest daughter is the lowest ranking, the next oldest daughter is next lowest in rank and so on. Baby females do not have a rank

▶ **School of snappers** (*Lutjanus* species). The reasons for group life are complex, but for these fish living in massive schools it probably reduces the chances of predation on each of them.

of their own but, as they approach puberty, they rise in rank above their elder sisters.

Rank pervades many features of Rhesus social life: it even affects the sex of a female's babies. High-ranking mothers give birth to more daughters than sons whereas low-ranking mothers do the reverse. This could have a valuable effect. The babies of low-ranking mothers will themselves be low ranking. If they are females, they are doomed to their low rank for life within the group into which they were born, but sons of low-ranking females will emigrate, and have an opportunity of rising in rank. So, the more sons to which a low-ranking mother gives birth, the greater will be the proportion of her offspring that have a chance of improving their rank. By contrast, the daughters of high-ranking females will be guaranteed a high rank when they grow up, whereas the sons of high-ranking females can at best only improve their rank a little when they emigrate, and may actually fall in rank. So, the more daughters to which a high-ranking female gives birth, the greater will be the proportion of her offspring that have a guaranteed high rank.

Social groups are constantly changing. Apart from births and deaths, older animals of one or both sexes are continually leaving their home group, and strangers are arriving. In many mammal species, individuals leave of their own accord but, in others, males in particular are driven out. Adult male lions evict adolescent males from their pride, and adult male squirrel monkeys, found in Central and South America, do the same to adolescent males that have grown up in their own group.

Sometimes, when adolescent males leave their home group, they join an all-male, bachelor band. This is very commonly the case in antelopes. Usually the resident males of the bachelor band do not resist the newcomer's entry. If, instead, a male, whether adolescent or adult, leaves its home group and tries to join another containing both adult males and adult females, the picture may be very different. An adult male Red deer will fiercely resist attempts by other males to come among his harem, and exhausting battles between them may follow. Japanese macaque males may, likewise, have to overcome attacks by one or more resident males if they are to join a new group.

Newcomers are not always treated with hostility, however. Adolescent male Rhesus monkeys may follow in the footsteps of elder brothers, and become gradually and peacefully assimilated into the groups that these older brothers had previously joined. Groups of Whip-tailed wallabies also seem not to behave aggressively to newly arrived males.

When a male leaves its group of birth, or its bachelor band, in an attempt to join a new group, it may face many dangers, not only from hostile rivals in the new group but also from predators while on its solitary journey. Nonetheless, if males do not make this journey, they may never be able to breed. In their group of birth, older males may prevent them from mating with females and, in bachelor bands, there are, of course, no females with which a male can mate. The disadvantages of transferring to a new group, though severe, may well be less, therefore, than those of staying put.

If animals are to live in a group, there has to be a way of keeping them together. Sometimes, the group contains a leader that the others follow—an arrangement found in cattle. In some species, the leader herds its group, and prevents them from wandering. In Hamadryas baboons, found in northeast Africa and Arabia, a male may have a harem of perhaps four females and their young. If any of the females wanders away from him, the male chases after her, bites her on the nape of the neck and drives her back to the harem. Many birds and mammals emit continual calls, and these become more frequent when the animals enter dense vegetation. These calls help the animals to keep in contact with one another. Animals may also stay in contact as a group by remaining attached to the same home area. Bats that return to the same roost each night and gulls that nest in the same colony stay together as

▲ **Bonding between mother and young**—a female Cape ground squirrel (*Xerus inaurus*) grooms her baby. Social interactions between individuals bind together animals living in groups.

▶ **Communal stomp.** Displays in large groups can be highly infectious, as exemplified by this group of Lesser flamingos (*Phoeniconaias minor*) parading around the feeding grounds.

▼ **Cooperative fishing.** White pelicans (*Pelecanus onocrotalus*) fish in groups, often encircling shoals of fish and preventing their escape.

groups largely as a result of their attachment to these sites.

Some of the most spectacular examples of group cohesion are found in schools of fishes. Hundreds or thousands of fishes in the school swim side by side in the same direction; they turn together, accelerate and slow down together, all in impressive synchrony. They do not rely simply on sight to achieve this: if fishes, such as saithe, have covers fitted temporarily over their eyes, they can continue to swim in synchrony with other members of the school. Indeed, their synchrony appears to be greater than normal as though, in their unusual condition, the fishes are unwilling to try out any behavior on their own. Fishes have structures along the sides of their bodies called "lateral line organs" that are especially sensitive to vibrations in the water. By means of these, they are able to pick up the eddies and currents made by the bodies of neighboring fishes in the school, and so are able to maintain a constant bearing and a minimum distance from them, even if they cannot see them.   NRC

DEVELOPMENT OF
SOCIAL RELATIONSHIPS

*The needs of a young animal. . . The parent–offspring bond. . .
Competition between siblings. . . Social play. . . Breaking the
bonds. . . Dispersal of young. . . Imprinting. . .*

IN many respects, the behavior of a young animal is less complete, less complicated and less competent than that of an
adult. The development of behavior is, in part, about the ways
that genes and conditions in the external environment work
together to elaborate and perfect behavior as the individual
grows up. It is often the case, however, that the needs of a
young animal are totally different from those of an adult. A
caterpillar not only looks different but behaves very differently
from a butterfly, and does so for good reasons because it lives
and feeds in such a different world. Similarly, the young mammal suckling from its mother feeds quite differently from the
way she does, again for good and obvious reasons. The development of social relationships reflects, in part, the changing
requirements of an animal as it grows up—from dependence
on parents, through acquiring skills as a juvenile, to competing
for mates and becoming a parent itself.

Of course, not every parent has a relationship with its offspring. The female herring releases her eggs into the sea where
some of them are fertilized by the sperm released by the male.
Neither parent has anything to do with the young. The same
is true in many other animals and, even in those that do show
parental care, a social relationship between parent and offspring may take time to develop. By degrees, the young animal
will usually come to play an increasingly active part in the relationships with its parents as it achieves greater mobility. It may
start to beg for food or do its part in keeping close to one or
both of its parents. A family party of shrews may, as it moves
around, form a caravan with the mother in front and her offspring strung out behind, each holding in its mouth the tail
of the one in front.

At this early stage, young mammals and birds could not
survive without parental care. The parent-offspring relationships are bound up with the giving and receiving of food, protection from danger and so on.

Parents quickly learn to identify their own offspring and so
reduce the risk that they will spend time and effort caring for
the young of others. Once this has happened, they may even
attack any young that they do not recognize as their own. It
is a common sight in spring to see a mother mallard duck driving off a lost duckling that tries to join her brood. It is not surprising, therefore, that, when parents behave like this, the
young should also quickly learn to identify one or both of their
parents. In a noisy colony of nesting terns it is remarkable how
a chick that has been dozing in its nest becomes alert as it hears
in the distance the call of one of its parents returning with a
fish. In mammals the male does not provide milk and the parent
with which the young one forms a strong relationship is
characteristically its mother. She provides its food and forms
a secure base from which it can start to explore the environment. A real mother provides both food and protection: she
is a haven for warmth and contact as well as a source of milk.
If models giving only food or only comfort are provided
separately in experiments, baby Rhesus monkeys resolve their

▲ ▼ **The needs of a young animal**
can be totally different from those of
an adult. A caterpillar not only
looks different, but, for example,
may feed on leaves ABOVE while the
adult sucks nectar BELOW. Shown
here are the caterpillar and adult of
the Large white butterfly (*Pieris
brassicae*).

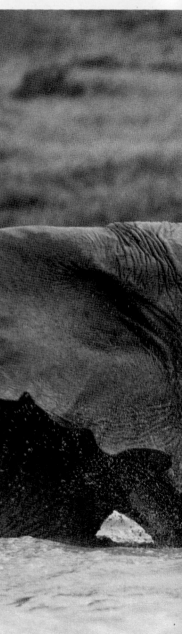

▶ **Playtime in the pool**—young
African elephants (*Loxodonta
africana*) frolicking in a water hole.
During play young animals may
acquire many social and
competitive skills that are useful in
adult life.

conflicting needs by feeding from the uncomfortable wire "mother" and, when frightened, clinging to the more comfortable cloth one, showing that this attachment is not simply to the source of nourishment.

If a young animal has siblings of the same age, the early social relationships between them are often competitive because the one that gets most care and attention from its parents grows fastest. Even at an early age, however, a nest full of animals, such as kittens, can help one another by huddling and thus conserving heat. As they get older and more mobile, cooperation becomes much more vigorous. The chance for social play is one way in which young animals may be individually aided by their contact with siblings and others of the same age.

In cats and monkeys, social play has been found to develop at a different time and in a different way from play with objects, and probably serves an exclusively social function. Several hours a day are spent vigorously chasing, mounting or mock fighting with siblings or others of the same age. Often, these playful activities are preceded by characteristic signals, such as that shown by dogs which drop down on their front legs with tails wagging. The bouts of activity can be intense and broken off as quickly as they started. It is very unusual for the participants in mock fighting to hurt one another, either with their teeth or, if they have them, their claws or horns.

Social play is most easy to recognize in young mammals. Whether it occurs more widely throughout the animal kingdom is a matter of dispute, because it is not wholly clear what play is for and, not knowing that, we cannot be sure whether this goal can only be attained by play or whether it is an objective that only the most complex animals have to reach. The presumption is, however, that playing is beneficial because the

◄ **The first solid meal** for a lion cub. Parents often force their young to take solid food earlier than the young would like. However, for the mother, this lessens the burden of suckling.

► **Babysitter and charge.** Gray meerkats (*Suricata suricatta*) live in small social groups. Often while the mother is foraging with the main group another adult will care for her young at the den. These babysitters are not necessarily related to the young.

▼ **Mistaken identity**—goslings follow a dog as if it was their mother. These goslings have never seen their true mother, but have been imprinted on the dog so treat it as if it is actually her.

Imprinting

When recently hatched birds, such as ducklings, are hand reared for a few days, they strongly prefer the company of their human keeper to that of their own species. This remarkable process which can so dramatically influence the development of social relationships is called "imprinting."

For many years, it was thought that this process enables an animal to learn about its species. It is becoming apparent, however, that imprinting serves a more subtle but equally important function enabling the young to recognize one or both of their parents as individuals. The "precocial" species that are feathered and active at hatching must learn the parental characteristics quickly because, on leaving the place where it was hatched, the young bird must stay near a parent. If it approaches another adult of its own species, it may be attacked or even killed. For this reason, as soon as the young animal is able to recognize its parent (or a substitute parent in an experiment), it escapes from anything that is noticeably different.

The rapid attachment to the first conspicuous object encountered after hatching, which almost invariably is the parent in natural conditions, and the fear generated by novelty after learning has taken place, means that imprinting usually occurs at a highly specific and early stage in the animal's life, the so-called "sensitive period." In "altricial" birds such as the swallow, which are hatched naked and helpless, learning occurs later in development and the young only respond selectively to their parents when they have left the nest about two weeks after hatching. In many birds, a form of imprinting occurs later still in development because, as they grow up, the birds come to recognize their siblings' adult appearance and use the information when they mate. The ideal mate must be a bit different but not too different from these close relatives to avoid the dangers of inbreeding and, at the other extreme, the dangers of mating with a genetically incompatible partner, such as a member of another species. As a result of having this mechanism for identifying close kin, strange things can happen in artificial conditions. Experiments have shown that, at a certain stage in its development, a bird exposed to the wrong object, such as another species, may end up with a strong sexual preference for that object.

animals involved in play can incur considerable costs. At the stage when social play is most common, considerable time and energy are devoted to it. By doing this, young animals frequently risk injury and make themselves more conspicuous to predators. Playful young cheetahs may even disrupt the hunting of their mother. It is exceedingly unlikely that such costs are incurred without some eventual benefits.

By playing with siblings or others of the same age, the young animal can acquire many social and competitive skills that could be useful to it as an adult. It could be more skillful in fighting and in mating. It could predict more easily what others are about to do and, in a general way, it may be more alert and more healthy than if it had no opportunities for play. None of this means, though, that such skills cannot be acquired in other ways, and play may be used as part of the basis on which adult behavior is built if the period of dependence on parents is prolonged. In such animals, play could be the most cost-effective way of acquiring the social skills required later in life.

Among the invertebrates, cooperation among close kin is seen most dramatically in social insects, such as ants and bees. Only close kin are allowed into the nest to reap the benefits of cooperation such as sharing the food. The identification of close kin once again requires a learning process somewhat similar to imprinting. One remarkable example of such learning is found in the Desert woodlouse in which up to 100 siblings live in an underground burrow. Individuals returning from foraging are allowed to enter the burrow because their odor is familiar. Unrelated members of the same species attempting to enter the burrow are not recognized, however, and may be killed and eaten.

Many animals have more than one brood of young in their lives and their reproductive interests are best served if each healthy offspring receives as much care and attention as the next one. When a young one can look after itself it may, therefore, benefit the parent to leave it to fend for itself and move on to the next brood. An offspring's long-term interests are best served, however, if it receives all the resources that its parents are capable of giving it. Thus, the interests of parent and offspring are not identical, and conflict arises between them. The conflict is particularly obvious in mammals in which the energy drain on the mother of providing milk for the young is severe. Mammalian mothers will often force their offspring on to solid food at a time when the young would quite readily go on suckling. The change can be seen in the increasing number of rejections by the mother when her offspring try to suckle. At a more subtle level, her role in staying near her young declines and they have to make more and more effort to keep in contact with her. In Rhesus monkeys the change takes place steadily over a period of months starting long before the time when the young monkey first takes solid food and continuing long past weaning. The mother may continue to provide protection and instruction for her growing offspring and, if the young one is female, it may stay with the mother even after sexual maturity. By this time, the relationship may be almost reversed with the daughter spending a considerable amount of time grooming her mother; and daughters may also help to care for their mother's subsequent offspring.

In most animals showing strong parental care, a time comes

when the bond with the parent is broken and the young move away. In general, male mammals move further away from their parents than females but, in birds, it is usually the females that travel the greater distance. Such dispersal usually takes place in young adolescents. It can occur quite suddenly and probably represents changes in the young as they mature rather than explicit rejections by the parents. After dispersal the young may attempt to find mates as soon as possible. In a polygamous mammal, such as a wild horse, however, the young males may form into bachelor herds until such time as they have grown enough to win a harem of females for themselves. When young male lions attempt to take over a pride of females from older males, they are much more likely to do so when they act as a group. Similarly, once in possession, they are better able to retain the pride when they cooperate with older males.

When viewed over the whole lifetime of an individual, social relations may change as old attachments are broken and new ones formed. In general, though, as a relationship develops between two individuals, it proceeds from a response to superficial characteristics in the other individual to a much more complex meshing of the behavior of the two with subtle and efficient communication between them. While it lasts, the cooperation involved in such a relationship is, in evolutionary terms, of mutual benefit to the participants and it pays each of them to interact with the other as efficiently as it can.

PPGB

ASSOCIATIONS BETWEEN SPECIES

Parasites—fleas, lice, worms... Slave-making ants...
Symbiosis—caterpillars and ants... Contract cleaners...

A PRIDE of lions or a hive of bees—these are just two examples from the many different kinds of associations that exist between individual animals of the same species. But a flea living on a dog's back, a sea anemone living on the shell of a hermit crab, or the bacteria living in the gut of a mammal are all types of associations between animals of completely different species which may or may not be to the benefit of both parties.

Most animal societies consist entirely of members of a single species, all of which gain from living in association with one another. There are, however, many cases where members of two species live together and these may be roughly split into three categories. First is "parasitism," where one species gains at the expense of the other, the parasite usually living either inside or on the surface of its host. Second is "commensalism"; in this, one species attends another and gains a living from it but without affecting it for good or ill in any way. The gulls that collect around sewage outflows and the foxes or raccoons that raid our garbage bins are good examples here. The third case is that of "symbiosis" or "mutualism" in which both parties benefit. We feed our dogs and they warn us of intruders. Of these three forms of association, parasitism and mutualism are the most common because if we look closely, the hosts of animals that appear to be commensals are usually affected in one way or another.

Animals can have many different sorts of parasites, living within them in such places as the gut, the blood or the liver, or living on their surface, feeding on their skin or sucking their blood. Most parasites tend to harm their hosts little because, if they damage them, the parasite's own chances of survival will be lowered along with those of the host. Instead, the way of life of the parasite is intricately matched to that of the host so that the two can live in relative harmony. A good example of such matching is in the rabbit flea which breeds in response to the hormone progesterone in the blood of its host; this hormone is a sign that the rabbit is pregnant so that there will shortly be another generation of hosts on to which the young fleas can climb. Getting from one host to the next, however, may involve more hostile tactics, especially in those parasites which must move between two different species. One flatworm which parasitizes fishes has to be transferred to a bird for its next stage. The chances of this are increased because the worms accumulate in the eyes of the fishes which, because their eyesight is impaired, swim more in the upper layers of the water making them more likely to be taken by gulls. There is even another flatworm which infects snails and which, when ready to transfer, enters their tentacles and makes them brightly striped so that predators are more likely to see them.

Parasites with behavior that is well adapted to their hosts are especially common among the social insects, where one species will often live in the nest of another. Usually, host and parasite are close relatives, presumably because the system relies on the parasite having similar signals to those of the host so that it is not evicted. In some ants, the queen invades the nest of another species, by speed, stealth or deception, and may then rely on her hosts for food or, more brutally, even kill the queens in the host nest so that the whole production line is moved over to rearing her own offspring. Another possibility is for ants to have nests of their own but to import livestock or labor, either aphids to provide them with honeydew or workers of other ant species to perform as slaves. Slave-making ants do have workers of their own, but the role of these is simply to form raiding parties to invade the nests of other species and carry off the pupae, the young from which will be their slaves.

Not all the behavior of ants seems so brutal and "uncivilized." They also provide some good examples of symbiosis. Some lycaenid butterflies lay their eggs near ant nests, and the caterpillars gain protection from predators and parasites from the ants. In return, they produce amino acids, chemicals on which the ants feed and which are essential to them if they are to produce the formic acid with which they protect themselves; amino acids are not present in the nectar on which they feed. Thus, both caterpillar and ant benefit—a fine example of a symbiotic relationship. In this case, one species gains protection and the other food, and these two benefits also

---

Contract Cleaners

There are many cases in the animal kingdom where one species helps to keep another clean, but this is not pure altruism; the cleaner is "paid" for its services with the food that it receives from the body surface of its client. The most famous example is the Crocodile bird, a plover which lives close to crocodiles and feeds from their surface. Although it is probably an exaggeration, it is even said that these birds can hop in and out of the mouth of a crocodile as it rests in the sun but remain unmolested because, like dental hygienists, they pry fragments of food from between the teeth of the reptile and so help to keep its mouth healthy.

Many species of fishes indulge in cleaning others, removing parasites from their outer surface, their mouths and their gills. The cleaners often stay in a particular area so that fishes needing to be cleaned just have to visit that place to receive their attentions. Cleaners tend to be clearly marked and to show striking behavior. For example, the cleaner wrasse approaches other fishes with a characteristic swimming movement, dipping up and down, which acts as a signal that it is a cleaner and should not be harmed. As in the case of the bird and the crocodile, the fishes to be cleaned are frequently large predators which normally eat fishes the size of the cleaners, so it pays the cleaner to advertise its occupation. The wrasse cleans the grouper and this species never harms it; when it is already thoroughly clean, the grouper swims away or hides to avoid the attentions of the wrasse. In some species, the client fishes have signals that indicate to the cleaners that they have had enough and are about to depart, such as shaking the body or rapid opening and closing of the mouth. The cleaners then swim out of the mouth or the gill covers, which have been specially spread to let them in, before the large fish closes them and swims off.

With such a beautiful and harmonious arrangement between two different species, it is not surprising that sometimes a third species has turned to it for its own benefit. There are two other fish species that mimic the wrasse in patterning and behavior, and can, therefore, approach close to fishes that these would normally clean. But they are predators and, when they get near, they dart in and bite off a piece of fin.

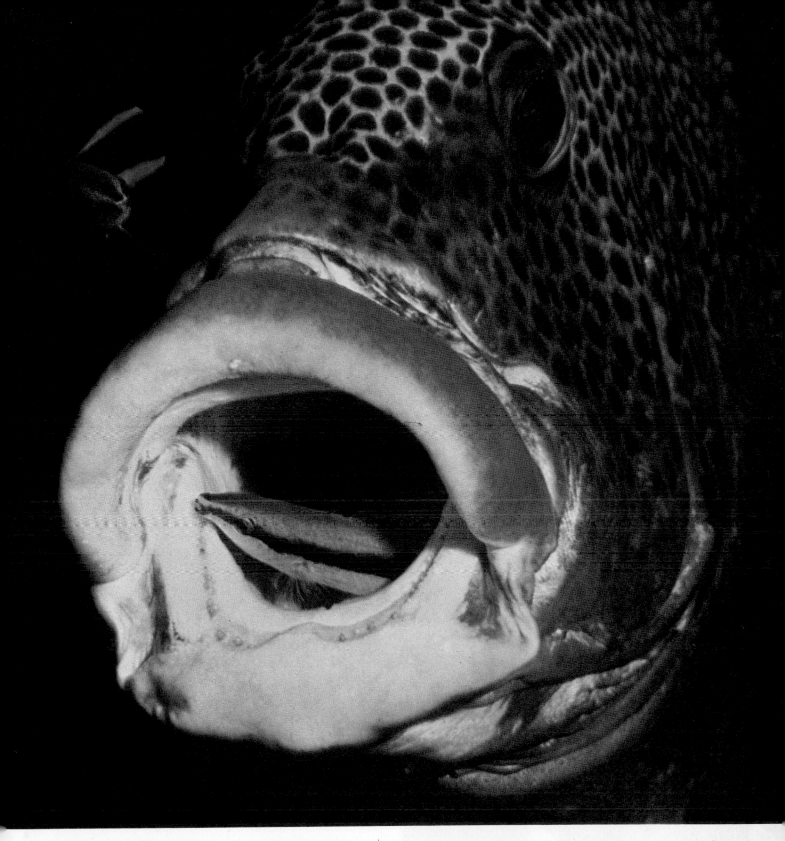

▲ **A dangerous way to live.** A monstrous sweetlips (*Plectorhynchus* species) relies on the attentions of the small cleaner wrasse (*Labroides dimidiatus*) to remove parasites from its mouth. On coral reefs, many fish, including predators, will attend so called "cleaning stations" where cleaner fish will remove parasites and damaged scales from their skin. Normal predator–prey relationships are not apparent at these times.

occur in many other cases where species associate with one another. For example, in the winter, birds such as thrushes, finches and chickadees often form mixed-species flocks, the individuals of which follow one another to food and also benefit because the flock has more pairs of eyes with which to spot predators. The stonechat, which is an especially vigilant bird, tends to collect a number of other bird species feeding around it and benefitting from its watchfulness, but this may be more a case of social parasitism than of symbiosis, for it is not clear that the stonechat itself gains from their presence.   PJBS

CULTURE IN ANIMALS

What is culture?... Information passed from individual to individual—memes... Foraging honeybees... Titmice feeding from milk bottles... Washing food by macaques... Finding nesting grounds in birds... Mobbing potential predators... Tameness and wildness... Cultural songs in saddlebacks... What sets humans apart?... Tool making... Language... Altruism... Human behavior and natural selection...

THE word "culture" means different things to different people. Some anthropologists prefer the term to be applied only to human beings largely because the array of customs and elaborate behavior patterns found in human societies is so rich and vast that they seem quite removed from anything found among animal societies. Others, including many biologists, seek a connection between human culture and behavior patterns in animals; they assume that an understanding of social animals will shed light on the origins of culture in humans.

With this in mind, it is helpful to define culture simply as "the transmission of information by behavioral means." By so doing, a sharp contrast is made with the transmission of genetic information where information can only be passed from parents to offspring. With behavioral transmission, on the other hand, any individual can pass information to any other individual. Richard Dawkins has called that which is exchanged a "meme." A meme is any bit of information, be it a fact, a fad, or rumor; it is purposely not defined too precisely so that the term can be as useful as possible and be easily compared with genes. Memes can be passed quickly, and can spread with great speed through a population, while the spread of genes is exceedingly slow and requires many generations. Another helpful way to stress the difference between the two is that genes can exist in organisms without memes, as in

EXAMPLES OF CULTURE

▲ **Milk bottle raiders**—a pair of Blue tits (*Parus caeruleus*) breakfasting on the best cream.

◄ **Potato-washing monkeys**—Japanese macaques (*Macaca fusca*) washing potatoes to remove sand.

Both the above behaviors are passed from individual to individual by learning.

► **Transmission of genes and memes.** (1) Genes can only be transmitted from one generation to the next; this takes considerable time, as breeding is necessary, and not all offspring within the breeding group will acquire the gene from their parents; genes cannot travel backwards through generations. (2) Memes can be transmitted by imitation and learning from one individual to any other, even unrelated ones. Meme transmission can be rapid, as it does not require breeding, and memes can pass backwards from offspring to parents, so long as all are living together. Some individuals may not acquire the meme. Related individuals are shown darker than unrelated ones.

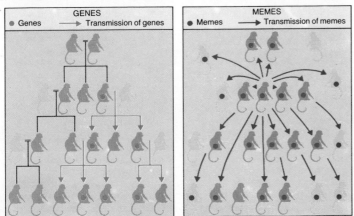

GENES	MEMES
● Genes ⟶ Transmission of genes	● Memes ⟶ Transmission of memes

plants, while it is impossible for memes to exist without genes. The genes are responsible for building the organism that is capable of exchanging memes with its kind.

If we use the simple definition of culture as the "passage of memes," we can begin to examine the evolution of culture; how did it arise, how did it develop into the remarkable culture that characterizes humans? The passing of behavioral information presupposes a nervous system and the equipment necessary to learn and possibly to teach. If we compare insects and vertebrates, we find that insects tend to respond to a signal by rigid responses while vertebrates have greater choice.

Among insects, the best-known example of transmission of behavioral information is found in honeybees. As Karl von Frisch showed, the scout bees can, through their dances in the hive, tell the other bees the direction of a source of nectar as well as its distance from the hive. The foraging bees register the message and rigidly follow the instructions (see p70).

If we turn to vertebrates, especially birds and mammals, patterns passed by cultural transmission may be of several sorts. First there are many examples where, either by imitation or by persistent teaching, young animals can learn how to obtain some inaccessible food. For instance, twirling a thin stick or straw down the hole of a termite nest and licking off the adhering insects require skill, and young chimpanzees will watch carefully as an older animal performs the task. Oystercatchers, common shore birds, have different sources of food depending upon their location. If they feed on marine worms and other soft invertebrates, a chick will stay with its parents for six to seven weeks; but, if mussels are the only readily available food source, the young will remain with their parents for 18 to 26 weeks. The reason for the extension in parental care is the great difficulty of learning how to open the mussel for its flesh.

The two most celebrated cases of the passing of behavioral information are particularly important because they involve an "innovation," an invention that spreads through a community. One is the discovery by some pioneer tits in England that, if the bird pecked through the cap of a milk bottle, it could drink the cream at the top. This trick spread from a single location by imitation so that now no milk bottle is safe anywhere in Britain. The other example is that of Imo, a young female Japanese macaque, that learned to wash the sand from sweet potatoes found on the beach, and later to do the same with wheat by throwing it on the surface of the water and skimming off the sand-free grains. Again both of these innovations spread rapidly through the troop to which Imo belonged.

Information about predators may also be spread culturally. Mobbing of hawks by birds is common and, in this way, they prevent a surprise attack and keep the predator on the defensive. Some elegant experiments have been carried out using two blackbirds in separate cages: one could see a stuffed hawk while the other could only see a stuffed harmless Australian honeyeater. The first bird mobbed the hawk and soon, by imitation, the second mobbed the honeyeater. Then the first bird and the hawk were removed so that a new, naive bird saw the second bird mobbing a honeyeater, and it soon learned to do the same. If we ask what the possible role of such behavior might be in nature, the answer may be that, by cultural transmission, a bird could quickly learn to identify new enemies but, had the transmission been genetic, it would have been far too slow to be effective.

Social relationships may be affected by cultural transmission between individuals or groups allowing the formation of bonds. Many species of birds perfect their songs or calls through hearing other individuals (see p102). In some small finches in which both sexes have contact calls and form monogamous pairs, they learn call structure from each other, so that the calls are the same within the pair and different from other pairs. This mutually shared call is probably important in maintaining the pair bond. In other cases, it is known that whole populations of birds will have their own song dialect, and various suggestions have been made to indicate why this is advantageous. Whatever the reason, these dialects are clearly learned and, therefore, culturally transmitted.

An interesting example of cultural transmission of song occurs in New Zealand saddlebacks which are birds the size of a blackbird. They are now very rare except on two small islands off the New Zealand coast. On one of these, a new population was established by the New Zealand Wildlife Service and has been studied for a number of years. The entire island population was small (fluctuating from 28 to 74 pairs) and it was possible to band all the birds so that individuals could be identified. Saddlebacks are highly territorial and form strong pair bonds that are apparently reinforced by the songs of the males.

As in some other species that learn features of their song in different geographic regions each group has its characteristic song structure or dialect. For other species, it has been postulated that the dialects are a means of keeping the populations separate and preventing them from interbreeding. Very much the opposite seems to be the case among saddlebacks. It has been found that they are unusual in that the males do not appear to learn new songs as fledglings, but do so later on in

Tameness and Wildness

When Charles Darwin visited the Galápagos Islands in 1835, he was struck by the extraordinary tameness of most of the animals, especially the birds. He reported that they could be approached closely enough "to be killed with a switch, and sometimes, as I myself tried, with a cap or hat." He also wrote that the ground birds had been even tamer at an earlier period before the islands had been much visited by humans. He quotes Cowley, who visited the Galápagos in 1684, saying, "Turtle doves were so tame, that they would often alight on our hats and arms, so that we could take them alive: they are not fearing man, until such time as some of our company did fire at them, whereby they were rendered more shy."

Is such shyness passed on from one individual to another, by cultural transmission, or does each animal need to receive a fright to produce wariness? An excellent case for cultural transmission comes from some observations of elephants by the Douglas-Hamiltons in the South African Park of Addo. The elephants were causing great damage to the neighboring citrus farms and, in 1919, a well-known hunter (with the rather splendid name of Pretorius) was hired to kill the entire herd of some 140 elephants. He killed the animals one by one and, consequently, the survivors became exceedingly dangerous to any intruding human and very wary, remaining in the dense bush during the day time. As a result, Pretorius could not find all the elephants; he

admitted defeat and the remaining 16 to 30 animals were fenced off in an area of about 3,200ha (8,000 acres). Even today, although no shooting has taken place for over 50 years, these elephants remain the most dangerous in Africa. As the Douglas-Hamiltons conclude, "Few if any of those shot at in 1919 can still be alive, so it seems that their defensive behavior has been transmitted to their offspring, now adult, and even to calves of the third and fourth generation, not one of which has suffered attack from man."

▶ **No fear of humans.** Sea lions (*Zalophus californianus*) swim happily with a bather around the Galápagos Islands. Tameness is a feature of island species where predators are few.

◀ **Mobbed kookaburra.** Mobbing of predators by smaller birds is believed to spread culturally. Laughing kookaburras (*Dacelo novaeguineae*) will eat young birds as well as snakes and lizards.

their youth. Furthermore, the young males tended to leave their natal territory and join a new one where there was a different dialect. On arrival in a new group, the wandering male rapidly learned the local song. Therefore, in this case, the different dialects do not seem to favor reproductive isolation between groups, but may be a mechanism to insure outbreeding, as the birds tend to nest in different areas from those from which they came.

Another important and interesting observation was made in a long-term study. Occasionally, a young male would make an error in learning the song and persist in his erroneous copy. Soon the "mutant" song would be imitated by others and, in this way, a new song form would arise by what has been called "cultural mutation." New dialects arising in this way are perpetuated by cultural transmission, a process very similar to many human cultural changes.

Finding a nesting region after a long migration has been assumed to be an example of cultural transmission where the young follow the parents on their first annual migration. That this is a correct assumption for some birds has been shown by detailed banding studies on snow geese. The parents and the young birds leave their nesting grounds at Hudson Bay and go to Texas for the winter. They remain together as a family and return to the same breeding colony next spring; the young birds have been shown the way and can now do it alone.

In all these examples of animal culture, we must ask if there is any special advantage over genetic transmission. Clearly, behaviors can be modified more rapidly, as in mobbing; and extremely detailed instructions can be passed on, as with migration routes and bird songs. These messages might be impossible to handle by genetic transmission; memes are far more flexible. Their very flexibility opened the gates during the early evolution of humans for the development of our own extraordinarily rich and variable cultures.   JTB

Is Man Different?

Human versus animal behavior

Before Darwin wrote *On the Origin of Species*, it was possible to view humans as quite different from animals and to think that the behavior of one was of no relevance to that of the other. But Darwin's theory changed this and made it clear that humans are a product of evolution like any other animal. Even so, it is still a matter of controversy whether studying the behavior of animals is relevant to that of humans, and also the extent to which differences in behavior between humans and other animals are of kind rather than of degree. In many of its aspects, human behavior is obviously much more sophisticated than that of other animals, for example, in language, in culture, in the making and use of tools, and in social relationships. It has often been argued that these features set humans apart; at the other extreme, it has been suggested that people are so much more aggressive than other animals that this is sufficient to indicate that humans must be viewed as quite separate from them.

Tool making and use are known to have occurred in people over two million years ago, and it is sometimes argued that this indicates that these people had become human. Today, we obviously depend heavily on many different types of tools. The shaping of stones into choppers and scrapers was a first step in the direction of the cars, cookers and computers with which we are now surrounded. That first step has its animal equivalents! Studies of many animal species in the wild have shown them to use tools. A famous example is the Woodpecker finch, from the Galápagos Islands, which uses a cactus spine to prise grubs from their holes in trees. In other cases, horses have been known to use sticks to scratch areas of their back which they could not reach otherwise, and chimpanzees have been seen to hurl branches at intruders as if they were spears and to use bundles of leaves to wipe dirt from their fur. In all these cases the objects used have been found lying around rather than specially fashioned, but there are also examples where animals show the rudiments of tool making. Not surprisingly, most cases concern apes: chimpanzees have been seen to tear down branches and strip bark from them so that they are suitable for extracting larvae from ants' nests; they can also form a mass of leaves into a loose sponge to extract water from holes they cannot reach with their mouths.

In language, too, some basic features are found among other animals, although there is a wide gulf between their capabilities and those of people. Some animal calls are rather like words, such as the alarm call a blackbird makes when a hawk appears and others rush for cover as though the bird had actually shouted the word "hawk." The skills of chimpanzees and gorillas trained to use sign languages and to interact with computers are further evidence that animals are capable of very complex forms of communication. It is not easy to find hard-and-fast rules which will draw a line between them and ourselves. Indeed, some features of human languages, which might seem clearly distinct from any animal signals, can have their equivalents in surprisingly simple animals. For example, one of our most subtle capabilities is that we can communicate with one another about subjects which are distant in time and space from where we are talking. But so can honeybees, for their dance tells others of the location of food they visited some time before, and furthermore it does it in a symbolic language, too!

Altruism involves cooperating with and helping other individuals without direct benefit to the altruistic individual. Humans certainly do this more than other animals, and giving gifts and offering assistance are widespread in human societies. "Helpers-at-the-nest" among birds and male baboons which assist one another to oust a dominant individual so that one of them can mate with his female are animal examples. In the animal cases, such assistance is only given to relations or to individuals that are likely to reciprocate later, as we would expect from evolution theory, but this is also probably true of the great majority of human examples. Again, we cannot say that humans are totally distinctive in this respect.

Animals fight a great deal and it is quite common for field ethologists, who spend many hours watching them, to see one kill another—much more so than would be the case if we watched the same number of humans for as long. In their relations with other individuals, humans are comparatively peaceable: it is only when they wage war that humans kill others in large numbers, but this is not "aggression" in the animal sense, with individuals displaying or wrestling with one another over food or territory. It is behavior which has emerged since humans separated themselves from the world to which natural selection adapted them and so, like many of the behavior patterns humans have developed in the past few thousand years, it is not something with a biological equivalent. Natural selection would not favor many things about human behavior, such as the adoption of unrelated children, the use of contraception, suicide, genocide and the pressing of the button that releases the nuclear holocaust. But we

▶ **Hammering at its meal**—a thrush at its anvil. Tool use is not confined to humans. Shown here is a Song thrush (*Turdus philomelos*) using a stone to break open the shell of a snail. Thrush "anvils" are used time and time again so that many empty shells become scattered around.

► **Tool use comparison** between man and other animals. (**1**) The hand axe fashioned from pieces of rock was probably the first general-purpose tool used by early man some 500,000 years ago. Some animals also use very simple tools, for example (**2**) chimpanzees use grass stems to extract termites from mounds and (**3**) Woodpecker finches probe holes in trees with a cactus spine to extract grubs. The leap forward for man lay in the continued development of tool technology which has resulted in the advanced equipment (**4**) we find around us today.

(**5**) This dart tip used by advanced hunting cultures in North America, some 37,000 to 10,000 years ago, was a step towards the use of (**6**) spears and harpoons for killing other animals for food. In the animal world (**7**) chimpanzees are similarly known to hurl branches at predators. (**8**) Whereas man once mainly fought his battles in the same way as animals, ie face to face and "man to man," with today's modern technology, warfare is often conducted (**9**) by men sitting in operation rooms or in complex machines of war, far removed from the site of conflict or battle.

have moved on to shape our own world, in some cases for good and in others for ill, and natural selection has little say in the matter.

Human behavior certainly appears to be much more complicated and sophisticated than that of even our closest relatives, the great apes. Yet these examples show that most of the differences are of degree rather than of kind, animals having the basis of many of the features we show ourselves. When detailed studies are made, we are often surprised by just how complex the behavior of animals can be. The really important differences between ourselves and other species have arisen since humans ceased to live in small groups in the African savanna following the life to which millions of years of natural selection had adapted us. In a short space of time, we have made our world very different from this and we must adapt our behavior to it for our biology cannot keep pace with such rapid change.

PJBS

Bibliography

The following list of titles indicates key reference works used in the preparation of this volume and those recommended for further reading.

Alcock, J. (1984) *Animal Behavior: An Evolutionary Approach* (3rd edn). Sinauer, Sunderland, Massachusetts.

Barnard, C.J. (1983) *Animal Behaviour: Ecology and Evolution*, Croom Helm, London.

Boakes, R.A. (1984) *From Darwin to Behaviourism: Psychology and the Minds of Animals*, Cambridge University Press, Cambridge.

Bonner, J.T. (1980) *The Evolution of Culture in Animals*, Princeton University Press, Princeton.

Brian, M.V. (1983) *Social Insects: Ecology and Behavioural Biology*, Chapman & Hall, London.

Bright, M. (1984) *Animal Language*, BBC Publications, London.

Broom, D.M. (1981) *Biology of Behaviour*, Cambridge University Press, Cambridge.

Brown, J.L (1975) *The Evolution of Behavior*, Norton, New York.

Catchpole, C.K. (1979) *Vocal Communication in Birds*, Edward Arnold, London.

Chalmers, N.R. (1979) *Social Behaviour in Primates*, Edward Arnold, London.

Daly, M and Wilson, M. (1983) *Sex, Evolution and Behavior* (2nd edn), Willard Grant, Boston.

Dawkins, R. (1977) *The Selfish Gene*, Oxford University Press, Oxford.

Dewsbury, D.A. (1978) *Comparative Animal Behavior*, McGraw-Hill, New York.

Edmunds, M. (1974) *Defence in Animals*, Longman, London.

Free, J.B. (1978) *The Social Organisation of Honeybees*, Edward Arnold, London.

Frisch, K. von (1966) *The Dancing Bees*, Methuen, London.

Gould, J.L. (1982) *Ethology*, Norton, New York.

Halliday, T.R. (1980) *Sexual Strategy*, Oxford University Press, Oxford.

Halliday, T.R. and Slater, P.J.B. (1983) *Animal Behaviour*, vols I-III, Blackwell Scientific Publications, Oxford.

Hansell, M.H. (1984) *Animal Architecture and Building Behaviour*, Longman, London.

Heinrich, B. (1979) *Bumblebee Economics*, Harvard University Press, Cambridge, Massachusetts.

Hinde, R.A. (1970) *Animal Behaviour*, McGraw-Hill, New York.

Hinde, R.A. (1972) *Non-verbal Communication*, Cambridge University Press, Cambridge.

Hinde, R.A. (1982) *Ethology*, Oxford University Press, Oxford.

Huntingford, F.A. (1984) *The Study of Animal Behaviour*, Chapman & Hall, London.

Krebs, J.R. and Davies, N.B. (1981) *An Introduction to Behavioural Ecology*, Blackwell Scientific Publications, Oxford.

Krebs, J.R. and Davies, N.B. (1984) *Behavioural Ecology. An Evolutionary Approach* (2nd edn), Blackwell Scientific Publications, Oxford.

Lea, S.E.G. (1984) *Instinct, Environment and Behaviour*, Methuen, London and New York.

Lorenz, K.Z. (1952) *King Solomon's Ring*, Methuen, London and Crowell, New York.

McFarland, D.J. (ed) (1981) *The Oxford Companion to Animal Behaviour*, Oxford University Press, Oxford.

McFarland, D.J. (1985) *Animal Behaviour*, Pitman, London.

Manning, A. (1979) *An Introduction to Animal Behaviour* (3rd edn), Edward Arnold, London.

Marler, P. and Hamilton, W.J. III (1966) *Mechanisms of Animal Behavior*, Wiley, New York and London.

Owen, D. (1980) *Camouflage and Mimicry*, Oxford University Press, Oxford.

Owen, J. (1980) *Feeding Strategy*, Oxford University Press, Oxford.

Schmidt-Koenig, K. (1979) *Avian Orientation and Navigation*, Academic Press, London and New York.

Sebeok, T.A. (ed) (1977) *How Animals Communicate*, Indiana University Press, Bloomington.

Slater, P.J.B. (1985) *An Introduction to Ethology*, Cambridge University Press, Cambridge.

Smith, W.J. (1977) *The Behavior of Communicating*, Harvard University Press, Cambridge, Massachusetts.

Tinbergen, N. (1958) *Curious Naturalists*, Country Life, London and Doubleday, New York.

Toates, F.M. (1980) *Animal Behaviour: A Systems Approach*, Wiley, Chichester.

Trivers, R. (1985) *Social Evolution*, Benjamin/Cummings, Menlo Park, California.

Wickler, W. (1968) *Mimicry in Plants and Animals*, Weidenfeld & Nicolson, London.

Wilson, E.O. (1971) *The Insect Societies*, Belknap Press, Harvard.

Wilson, E.O. (1975) *Sociobiology: The New Synthesis*, Belknap Press, Harvard.

Wittenberger, J.F. (1981) *Animal Social Behavior*, Duxbury Press, Boston.

Picture Acknowledgements

Key: *t* top. *b* bottom. *c* center. *l* left. *r* right.
Abbreviations: A Ardea. AN Agence Nature. ANT Australasian Nature Transparencies. BCL Bruce Coleman Ltd. FL Frank Lane Agency. NHPA Natural History Photographic Agency. OSF Oxford Scientific Films. P Premaphotos Wildlife/K. Preston Mafham, PEP Planet Earth Pictures/Seaphot. SAL Survival Anglia Ltd.

1 P. 2 M. Fogden. 3*t* A. 3*b* PEP. 4 A. 5 SAL. 6*l* BBC Hulton Picture Library. 6*r* Mansell Collection. 7*t* NHPA. 7*b* Associated Press. 8 Swift Picture Library. 9 AN. 11 BCL. 12–13 P. 14 Jacana. 15*t* PEP. 15*b* P. 18–19 OSF. 20 P. 21*t* M. Fogden. 21*b* OSF. 24–25 BCL. 25*b* P. 26–27 Eric and David Hosking. 28–29 BCL. 31 P. 32 A. 33*t* BCL. 33*b*, 34–35 P. 36*b* BCL. 36*t*, 36–37 M. Fogden. 38 P. 39 M. Fogden. 41 P. 42 M. Fogden. 43 NHPA. 44 BCL. 44–45 ANT. 47 Eric and David Hosking. 48–49, 50–51 SAL. 52 A. 52–53 Biofotos/Heather Angel. 56–57 Frans Lanting. 57*t* BCL. 60–61 Nature Photographers. 62*t*, 62*c* P. 62*b* A. 63 BCL. 64 Prince & Pearson. 65, 66 Jacana. 67*t* M. Fogden. 67*b* David Hosking. 68–69 BCL. 69*b* FL. 71 OSF. 72–73 G. Frame. 76–77 BCL. 78*t* N.

Bonner. 78*b* Dwight R. Kuhn. 79*b* FL. 80–81 BCL. 84 R. Pellew. 85 P. 87 BCL. 88 M. Fogden. 89 NHPA. 92–93 M. Fogden. 93*b* NHPA. 94–95 BCL. 97 OSF. 98, 99*tl*, 99*cl*, 99*bl* T. Roper. 99*r* Frans Lanting. 100 BCL. 100–101 PEP. 102 Dwight R. Kuhn. 104 Eric and David Hosking. 106–107 M. Fogden. 108–109 BCL. 111*l* OSF. 111*r* C.A. Henley. 112–113 BCL. 113 Fred Bruemmer. 114 PEP. 115 A. 116–117 Biofotos/Heather Angel. 118 P. 118–119 A. 120–121, 122 NHPA. 123 P. 126–127 SAL. 127*b* Frans Lanting. 128–129 PEP. 130*t* SAL. 130–131 Tom Owen Edmunds. 131 J.B. Davidson. 132*t*, 132*c*, P. 132–133 NHPA. 134*t* PEP. 135 D.W. Macdonald. 137 PEP. 138*b*, 138–139 BCL. 140, 140–141 SAL. 142–143 OSF.

Artwork

All artwork Oxford Illustrators Limited unless stated otherwise below.
Abbreviations: PB © Priscilla Barrett. SD Simon Driver. RG Robert Gillmor. RL Richard Lewington. DO Denys Ovenden.

4 RG. 9 SD. 10 PB. 15 SD. 16, 17, 22, 23 DO. 24, 28, 29 PB. 30 RL. 38, 40, 43, 44, 46, 49, 52 SD. 54, 55*t* Ian Willis. 55*c*, 57, 58*b*, 59*b*, SD. 58*t*, 59*t*, Mick Loates. 64, 68 PB. 70*t*, 71 RL. 70*b* SD. 73, 74, 75, 79 PB. 82, 83 DO. 86 PB. 88 RL. 89 SD. 90, 91, 96, 97 PB. 103 SD. 104, 105*t* RG. 105*c* SD. 108 RG. 110 SD. 111 RG. 113 SD. 114 RG. 119, 120 RL. 123 SD. 124, 125 PB. 127, 139 SD. 143 PB.

GLOSSARY

Aestivation dormancy during the summer, equivalent to HIBERNATION in the winter. Mainly found in species living in areas, such as deserts, where drought and very high temperatures make the environment hostile. By becoming torpid in an enclosed and humid burrow, such animals can conserve both energy and water.

Aggression this is a controversial term, best used as a loose categorization of attack and threat behavior but sometimes taken to include a much broader group of activities. A cat attacking a mouse (**predatory aggression**), a bird singing (a behavior that tends to repel rivals) and a human speaking assertively, have all sometimes been described as aggressive. As the causes of these actions probably have little in common with fighting within a species, such broad usage is probably best avoided.

Agonistic behavior any behavior pattern associated with fighting and retreat, such as attack, escape, THREAT, defense and APPEASEMENT.

Allogrooming mutual grooming between animals, especially common among primates and birds. Allogrooming may assist in the care of the body surface because it is often directed to areas the individual cannot reach. Most often it is shown to mates, social partners and dominant individuals, however, and certainly also has the social function of cementing relationships between individuals.

Altricial describing those species in which the offspring are relatively helpless at birth and are kept for a period thereafter in a nest or other hiding place (cf PRECOCIAL).

Altruism conferring a benefit on another individual at a cost to the individual behaving altruistically. In this general sense, altruism includes sharing food and producing warning calls. In these cases, however, the altruist may benefit if the help given is to relatives (as they share a proportion of the altruist's genes) or to social partners who will reciprocate later. Most apparent altruism comes into these categories and so, strictly, is not altruism at all.

Amplexus the posture adopted by pairs of frogs and toads during the mating season, with the male riding on the back of the female.

Anisogamy the situation where germ cells (gametes) differ is size, as found in all higher animals, the female producing relatively few large eggs and the male many times that number of much smaller sperm.

Antithesis a principle proposed by Darwin that animal signals with opposite MESSAGES were also opposite in form. For instance, he contrasted a hostile dog, with erect posture and tail and head held high, with a friendly one, crouching low with tail between its legs.

Appeasement behavior which inhibits attack in situations where animals cannot escape or, as during courtship, where it is disadvantageous for them to do so. Often opposite in form (see ANTITHESIS), as well as in the message conveyed, from THREAT displays. Thus, gulls hide their beaks when appeasing, show them in threat. Other appeasing postures derive from juvenile behavior and, in monkeys, from the sexual presentation posture of females (see LORDOSIS).

Appetitive behavior the active searching behavior shown by animals seeking a goal, and contrasted with the CONSUMMATORY behavior they show when they reach it. When hungry or thirsty, a rat will explore actively until it finds food or water. Females of many species also become more active when on heat and this makes them more likely to come into contact with males.

Assortative mating mating in which an individual chooses a mate non-randomly in relation to its own characteristics. If **positive**, the mate chosen is like itself, if **negative**, it is unlike itself.

Attention state in which an animal is more responsive to one aspect of its environment that to others. Experiments have shown that animals may attend to and learn about some features of a situation and ignore others which are less relevant to them.

Bait shyness failure of pest animals to eat poison put out for them. This may stem partly from NEOPHOBIA and partly from the avoidance animals show for foods that have previously made them sick.

Behaviorism school of psychology founded by J. B. Watson which rejected introspection and stressed the importance of objective observation. Their studies were largely concerned with experiments on learning carried out in highly controlled and simplified laboratory situations.

Bioluminescence the production of light by plants and animals. Bioluminescence may be used in visual displays or, as in some deep-sea fish, to help in food searching in dark surroundings.

Brood parasitism the system in which the young are raised by individuals that are not their parents. Usually, as in the Brown-headed cowbird or European cuckoo, the hosts are of a different species and all young of the parasite species are raised in this way. Brood parasitism may also occur within a species, some females laying eggs in the nests of others.

Camouflage color and patterns of animals which blend with their background so minimizing risk of predation or, in the case of predators, enabling undetected approach to prey.

Cannibalism eating members of an animal's own species. It has been reported in many species but is relatively uncommon, except in animals subjected to stress or high population density.

Carnivore any mainly or exclusively meat-eating organism; alternatively, a member of the order Carnivora, many of whose members are carnivores.

Caste individual members of a species which are adapted to perform a particular task both by their structure and their behavior. Usually applied only to social insects such as honeybees and termites.

Causation describing the immediate causes of behavior and the mechanisms by which they act. Studies of causation investigate factors such as how external stimuli and internal factors, such as hormones, affect behavior and the neural structures involved in its production.

Chaining sequences of behavior in which actions occur in a series, each one leading to the next. Within an animal, this may be because one activity brings the animal into the situation where the next is stimulated. Chaining may also occur between animals, for example in courtship, where each partner often responds to the previous action of the partner in a series that leads to mating.

Cheating behavior believed to have evolved because the animal showing it gains at the expense of other individuals. For example, male scorpion flies present prey to females before mating, but some males cheat by mimicking females, flying off after accepting the gift and using it in their own courtship.

Circadian rhythm a rhythm which is about one day in length, found in many aspects of behavior and physiology. Circadian rhythms often appear even in animals kept under constant conditions but they then usually "free run" at a length which diverges from 24 hours. In nature this intrinsic rhythm is entrained by the cycle of light and dark and so matched precisely to the 24-hour rhythm of the earth's rotation.

Comfort movements a varied group of activities including grooming, scratching, shaking, stretching and yawning.

Commensalism an association between two species such that one of them benefits without appreciable cost to the other. Examples are SCAVENGERS, and predatory fish that live in shoals of harmless ones so gaining CAMOUFLAGE from which they can attack prey.

Conditioning learning by association. In classical conditioning two stimuli are associated so that the second comes to elicit a response formerly only elicited by the first. In instrumental or OPERANT conditioning, the rate of a response is raised or lowered by its association with REINFORCEMENT.

Conflict an animal with an approximately equal tendency to do two things at once is said to be in a MOTIVATIONAL conflict. Thus, a hungry animal beside whose food a frightening object is placed is in an approach-avoidance conflict. In such a balanced state, animals often show other, seemingly irrelevant, behavior (see DISPLACEMENT ACTIVITY). Many courtship displays are thought to have arisen from such actions produced as a result of conflict.

Consort relationship a temporary PAIR BOND between a male and a receptive female as found, for example, in Yellow baboons.

Consummatory act an act that terminates a behavior sequence and leads to a period of quiescence. Opposed to the APPETITIVE BEHAVIOR that leads up to it: thus, food seeking ends in eating, and mating in ejaculation.

Contact call any sound which serves to keep members of a pair or social group in touch with one another. Contact calls are commonly produced by moving animals especially where visibility is poor.

Cooperation the situation in which animals assist one another for mutual benefit. Thus, animals in groups may assist one another in predator detection because they have more pairs of eyes. They may also be able to catch larger prey through cooperative hunting, as in hyenas.

Coprophagy eating of feces. This is important in some animals to transmit organisms, which help to break down food, from one individual to another, such as from adult to offspring. Eating its own feces may also improve the efficiency with which an animal utilizes its food.

Countershading an aspect of CAMOUFLAGE in which an animal is paler colored in less brightly lit areas, usually the underside, so that it appears to be uniform and is less easily detected.

Coyness reluctance to mate often found in newly formed pairs. In females coyness may serve to ensure that the male is fully committed to her and will assist in care of young; in males it may ensure that young are his rather than of any previous male.

Crepuscular describing animal species that are active at dawn or dusk or both. Desert animals are commonly crepuscular because hot and dry days and cold and dark nights are hostile.

Crypsis see CAMOUFLAGE.

Cuckoldry the situation in which the female of a pair mates with males other than her partner. It is not infrequently observed, and mate guarding by males is thought to have evolved as a device to avoid it and the possibility of the male rearing young which are not his own.

Cultural evolution changing behavior as it passes from one generation to another through learning. Single individuals may learn new habits and, if these are copied by others, the change may spread and persist in the population without any genetic alteration being involved.

Culture a set of patterns of behavior which are reproduced generation after generation through learning, as well as the products of those behavior patterns, such as tools and works of art in humans (see also TRADITION).

Deceit see CHEATING.

Desertion the situation in which one partner leaves parental care to the other at some stage after mating. Whole species may show the same pattern: eg female alone caring (ducks) or male alone (sticklebacks). In other cases, desertion occurs occasionally where the deserted partner can rear young alone and the deserter can find a new mate. The term is also applied to desertion of offspring by adults which may occur after disturbance or in adverse conditions.

Dialect variation of social signals within a species from one locality to another. Described for vocalizations of birds and frogs, and for the waggle dance of the honeybee. The term is sometimes restricted to cases where variation shows sharp boundaries between populations giving a mosaic of signal forms.

Discrimination the ability of an animal to distinguish between two stimuli. Thus, gulls prefer to sit on brown egg models rather than on red ones and so must be able to distinguish between them. Many psychological studies have examined the processes of discrimination learning whereby animals can be trained to make such distinctions.

Displacement activity seemingly irrelevant behavior, such as grooming or nest building, sometimes shown by courting or fighting animals. It is thought to be shown where more relevant behaviors are THWARTED or in CONFLICT. Such actions are often hurried or incomplete and some are considered to have been the original behavior from which certain DISPLAYS evolved.

Display movement pattern used in communication. These are often striking, stereotyped and species-specific, especially during courtship and aggressive behavior.

Displays may also function between species (see also DISTRACTION DISPLAY).

Disruptive coloration color patterns which break up the outline of an animal and other features of its body so that it is very hard to pick out when motionless.

Distraction display a DISPLAY with which mother birds lead predators away from their broods by feigning injury.

Dominance the situation in which one animal dominates another in fights or in access to resources such as sitting positions, food and mates. Dominants are often older or stronger, but fighting may not be involved in encounters because subordinates frequently defer. In some social groups relationships may be consistent and clear enough for animals to be placed in a linear dominance hierarchy in which each is dominated by those above it and dominates those below.

Drive psychological term applied to supposed moving forces underlying the appearance of behavior. For example hunger drive, sex drive and social drive. The term has fallen from use as the causes of behavior have been analyzed in more detail and different factors have been found to affect different aspects of behavior without the need for motive forces affecting each of them to be postulated.

Duetting the calling of some pairs of birds, notably African shrikes, where calls of two members of the pair are so close together and coordinated that the sequence sounds like the song of a single bird. Duetting may help to maintain the pair bond as well as having a function in territorial advertisement.

Echolocation orientation by emitting high-pitched sounds and locating the positions of objects by the way in which they reflect the sound. Echolocation is used mainly by bats, but also by dolphins and oilbirds.

Electrocommunication communication by means of electric fields and pulses generated by one individual and detected by others. Electrocommunication is used by some fish in courtship and aggressive encounters.

Emancipation a suggested process whereby movements shown during DISPLAYS have become divorced from their original controlling mechanisms during RITUALIZATION. Thus, grooming and drinking movements acting as signals during courtship may no longer require MOTIVATION to groom or to drink before they appear.

Engram the memory trace within the brain, the nature of which is as yet unknown. Many electrical, chemical and anatomical studies of the brain are aimed at discovering its form.

Epigamic selection sexual selection based on the choice of one sex by the other, usually of males by females, which has led to enormously elaborated DISPLAYS. Such courtship displays are often called **epigamic displays**. The classic example is the peacock's tail.

Estrus the state in female mammals when they are receptive to a male or "on heat."

Ethogram an inventory which lists and describes all the bahavior patterns shown by a species.

Ethology the biological study of behavior.

Eusociality the form of society found in some social insects in which there is cooperation between individuals and division of labor. For example, some individuals are more or less sterile and others primarily engaged in reproduction. (See CASTE.)

Evolutionarily stable strategy see STRATEGY.

Extinction the process whereby learned behavior ceases to be performed when no longer appropriate. In a rat the response of bar pressing to obtain food will slowly decline when the reward is withdrawn; dogs will cease to salivate to the bell announcing food if it is often rung without food appearing.

Feedback the modification of behavior in response to its consequences. In **negative feedback** the consequences slowly suppress the behavior, as when grooming removes the irritation that caused it. **Positive feedback** has the opposite effect leading to more of the behavior, for example where sexual sitmulation leads to more sexual behavior and so on until terminated by ejaculation.

Filter feeding a method of feeding used by some animals whereby they filter small particles from the surrounding medium as they move through it, as in whales, or as it moves past them, as in barnacles.

Fitness a measure of an individual's genetic contribution to the next generation. The term **inclusive fitness** is used to refer not only to the fitness an individual achieves through its own reproductive efforts but also the improved fitness it achieves through assisting relatives to reproduce. Animals are thought to behave in such a way as to maximize their inclusive fitness.

Fixed action pattern any stereotyped movement pattern found in more or less identical form throughout a species, at least among animals of similar age and sex. Their relative fixity makes such movements especially useful in classifying animals and in analyzing behavior. As they are not as fixed as originally thought, the term **modal action pattern** has been proposed as preferable.

Function the function of an attribute, such as a behavior pattern, is its selective advantage. This is the way in which it confers on its possessors greater survival and reproductive success than if they did not have it, and thereby persists in the population.

Goal-oriented behavior behavior directed towards some goal which, when reached, brings that behavior to an end. It is most obvious in, for example, the case of a hungry rat running a maze to reach food, but can be applied to a variety of behavior patterns, such as mate seeking leading to copulation, and nest building leading to a nest.

Graded signal animal signals which show differences in frequency or intensity depending on the state of the individual showing them, just as people shout louder the more angry they are. Often, animal signals show TYPICAL INTENSITY.

Gustation the sense of taste. Gustatory stimuli are those received through this sense.

Habitat the external environment to which an animal species is adapted and in which it prefers to live, defined in terms of such factors as vegetation, climate and altitude.

Habituation a simple form of learning whereby an animal ceases to respond to a stimulus presented to it repeatedly but which is neither noxious nor rewarding. A property of the nervous system, it is distinguished from SENSORY ADAPTATION and muscular fatigue; these may lead to similar results but usually with more rapid recovery when the stimulus is witheld.

Haplodiploidy the genetic system of some insects, notably ants and bees, in which males develop from unfertilized eggs and females from fertilized ones. Thus, males are haploid, having half the number of chromosomes possessed by the diploid females.

Harem a group of females guarded by a single male that mates with them and drives off other males attempting to do so. Examples are in Red deer and in elephant seals.

Helper an individual that assists in the rearing of offspring which are not its own, shown by many bird species. The helpers are usually elder siblings of the brood.

Herbivore an animal which eats mainly plants or parts of plants.

Heritability a measure of the extent to which variation in a behavior pattern, or other characteristic, is due to genetic rather than environmental causes. Only when heritability is high is selective breeding likely to alter a trait.

Hibernation adoption of a state of dormancy in the winter such that heart rate and temperature fall, and hence energy requirements are reduced to a minimum. Hibernation enables animals, such as squirrels, bats and hedgehogs, to survive periods of hostile climate and low food availability.

Hierarchy see DOMINANCE.

Hoarding storing of food for later use either in a cache or, as in **scatter hoarding,** with each item in a separate place. Mammals, such as foxes and squirrels, and birds, such as Marsh tits and Acorn woodpeckers, show hoarding behavior.

Home range the area which an animal or group of animals occupies or visits. As it is not necessarily defended from others, it is distinguished from TERRITORY.

Homing the ability displayed by many species to return to the same place, usually a breeding area, from a distance, either as part of their normal activities or after displacement in an experiment. Good examples are fish, such as eels and salmon, turtles and homing pigeons.

Imitation the copying of the behavior of one individual by another such that the second acquires a new behavior pattern. Some behavior, such as patterns of bird song, may pass from generation to generation in this way, and novel forms of behavior discovered by one animal may spread rapidly through the population by imitation (see also CULTURAL EVOLUTION).

Imprinting a process whereby young animals learn the characteristics of other individuals, normally a parent, early in life. In **filial imprinting** they come to devote their social responses to that individual. Learning about parents and siblings may also influence mate choice through **sexual imprinting,** the animal seeking a partner similar to those with which it was reared, but not usually identical with them.

Incentive characteristic of a stimulus which makes it pleasant or unpleasant to an animal. Thus, to a rat, sweet solutions have high incentive value and bitter ones have low incentive value.

Incest the mating of close relatives, usually of siblings or of parents with offspring. It is seldom found among animals partly because the young of one sex or other tend to disperse but partly also because individuals tend to prefer less closely related partners.

Innate behavior a term which has suffered from a variety of meanings and has, as a result, largely fallen from use. It could be taken to imply that a behavior developed without learning, without practice, without copying from others or even without any environmental influence at all. Criticisms have been especially directed at this last sense with its suggestion of inflexibility and genetic determinism.

Insight learning which involves the appreciation of complex relationships and, at its most sophisticated, may imply thought and reasoning. It is not clearly distinguished from other forms of learning and its applicability to animal examples is doubtful.

Instinct a term seldom used now but with many meanings in the past. Most often, it was applied to systems of behavior, like the DRIVES of psychologists, which were thought by ETHOLOGISTS to be inborn and fixed, for example reproductive instinct. The term **instinctive behavior** was used in the same way as INNATE BEHAVIOR and has fallen from use for similar reasons.

Intelligence that capacity which enables an individual to learn tasks, reason and solve problems. Such capabilities being based on many attributes, testing of intelligence in humans is open to numerous biases. In animals, intelligence is also hard to assess: they may find superficially difficult tasks to which they are adapted easy and other, apparently simpler, tasks much more difficult.

Intention movements movements shown by an animal just before commencing an activity which indicate to an observer what it is about to do. For example, many birds crouch down in a hunched posture before taking flight. In many cases such actions are thought to have evolved into signals as they often occur during courtship and aggressive encounters and so indicate the animal's intentions to its partner or rival.

Kaspar Hauser a boy found in Nuremberg in 1828 who behaved as a child and later claimed to have been raised by a man in total isolation. He has given his name to the **Kaspar Hauser** or **deprivation experiment** in which animals are reared in isolation and the effect of this on the development of their behavior is noted.

Kinesis movement of an animal which is affected by a stimulus but not oriented with respect to it. Thus, animals that turn less or move more quickly in the light tend to accumulate in darker places without the direction of the light necessarily affecting them (cf TAXIS).

Kleptoparasitism the habit of some animals, such as the parasitic jaeger, of living on food stolen from other species.

Language a term usually applied only to the verbal communication system of humans but sometimes also applied to aspects of animal communication. That language in humans can be sharply distinguished from all animal signals has become increasingly doubtful and an elaborate definition is required if it is to exclude all animal examples.

Lek a communal mating ground, used especially by many bird species and a few mammals, where males set up and defend small TERRITORIES packed close together in an arena. Females visit the area for mating and the sole function of the territories seems to be in female attraction.

Lordosis the posture shown by receptive females of many mammal species in the presence of a male. The back is curved downwards so that the hind quarters are raised, and the tail is deflected to one side making it easy for the male to mount. The equivalent posture in primates, **presentation**, has become a social signal used by both females and males when submissive.

Meaning the information gleaned by the recipient of a signal. This can only be assessed in animals by the way in which it responds to the signal. To an unmated female the meaning of a male courtship display may be "approach and attempt pair formation" (cf MESSAGE).

Message the information about the sender encoded in a signal. Thus, a male courtship display may convey the message "I am an unmated male in breeding condition" (cf MEANING).

Migration the long-distance movements of animals and especially the regular seasonal movements between breeding and wintering grounds or different feeding places shown by many birds and fish as well as some mammals and insects.

Mimicry resemblance between two animals in behavior or physical features such that the two are confused to the advantage of one or both of them. In **Batesian mimicry** an edible mimic resembles a distasteful **model** making predators less likely to eat it. In **Müllerian mimicry** features are shared by two or more noxious species so predators learn to avoid all of them more easily. In **aggressive mimicry** a predator resembles a harmless or attractive species enabling it to approach prey more closely. The term mimicry may also be applied to the IMITATION of behavior as in birds which learn the songs of other species.

Mobbing the response shown by some animals, especially small birds, to potential predators, such as owls, whereby they approach the predator and fly around it calling noisily. Several possible functions have been suggested for this: for example it may drive off the predator or draw the attention of others to the danger.

Modal action pattern see FIXED ACTION PATTERN.

Motivation the internal changes responsible for making an animal behave differently at different times. Confronted with food, animals do not always eat, nor do they attempt to mate whenever a potential partner appears. The study of motivation is aimed at discovering the processes responsible for such changes.

Mounting the posture, on the back of the female, adopted by the male during mating in most birds and mammals.

Mouthbrooding a mode of parental care shown by some fish in which the female protects the eggs within her mouth until they hatch; the young may also return to the mouth thereafter when danger threatens.

Mutualism association between different species to their mutual advantage (see also SYMBIOSIS).

Navigation the ability of animals to find their way to a goal regardless of the location from which they start (see also HOMING). To do this requires more than just ORIENTATION using a "compass," but also information equivalent to possessing a map, as the compass direction of the goal depends on knowing the position of the starting point.

Neophobia dislike of novelty, especially where animals reject food which is novel or presented to them in a different place (see also BAIT SHYNESS).

Niche the position in the ecological community occupied by a particular species defined by the HABITAT it occupies, what it eats and what it is eaten by.

Omnivore an animal which eats a varied diet including both animal and vegetable matter.

Ontogeny the development of an individual organism. The study of ontogeny examines the way in which genetic and environmental factors mold behavior during the lifetime of the individual animal.

Open field an arena, usually divided into squares, used to study the activity and exploration of animals.

Operant a term used by Skinnerian psychologists to denote a behavior pattern that could be modified in some way by its consequences. For example, lever pressing is an operant because its frequency rises if it is followed by delivery of food to a hungry animal.

Optimal foraging the theory that animals searching for food should behave in the most efficient way. This is most often taken to imply that they should behave so as to maximize their net rate of energy intake, that is, the energy they gain less the energy they expend in each unit of time spent foraging. Experiments suggest that many animals come remarkably close to achieving this.

Optomotor response the response whereby animals attempt to keep themselves static in relation to their visual world. Thus, a wide variety of animals placed in a striped drum that rotates slowly will move their head and eyes to follow the stripes and keep the image still.

Orientation the movement of animals in relation to their environment. For example, animals may orient in relation to landmarks or some goal that they can perceive. At a more complex level, some animals are known to use the sun, stars or the earth's magnetic field to orient in a particular compass direction (see also NAVIGATION).

Pair bond the attachment between members of a mated pair causing them to stay together. In some species, such bonds may persist throughout life but, in others, they last only for one breeding season or mating period (see CONSORT RELATIONSHIP). In POLYGAMOUS species one individual may have pair bonds with several others at the same time.

Parasitism the system in which one species gains a living at the expense of another, usually either within it or feeding from its surface or blood. In **social parasitism** one species is dependent on the societies of another, living in them and gaining food from them (see also BROOD PARASITISM).

Parental investment any investment by a parent in an individual offspring which increases the offspring's chances of surviving at a cost to the parent's ability to invest in other offspring. On the basis of this theory parents should not invest too much time and effort in one offspring but should allocate their investment to maximize the number of their young that survive to breed.

Parturition the act of giving birth.

Peck order an alternative term for a DOMINANCE hierarchy, usually only applied to birds. This phenomenon was first described in flocks of chickens.

Pheromone a chemical substance which passes through the environment from one animal to another, either as a signal which elicits a behavioral reaction in a recipient or as a messenger leading the recipient to show a physiological change. Most examples of the latter concern reproductive effects, such as faster maturation brought about by the smell of the opposite sex.

Play perhaps the most difficult category of behavior to define, because it involves many types of behavior shown in other contexts, such as fighting and prey capture. Play often lacks the organization, completeness and "earnestness" of the same behavior in other contexts. Sometimes split into **object play** and **social play**, it is almost entirely restricted to mammals and most common in young ones. Its functions have been the subject of heated and unresolved debate.

Polygamy the mating system in which one individual has two or more mates, either simultaneously or successively (the latter is also known as **serial monogamy**). In **polygyny** one male has several females. In **polyandry** one female has several males. The latter is rare because the reproductive success of females is usually limited by the number of eggs they can produce rather than by the number of mates they can have.

Precocial describing those species in which offspring are able to move about and feed themselves at an early age. Unlike ALTRICIAL species, these young are well developed at birth, with fur or feathers and open eyes, and they start to walk soon afterwards.

Presentation see LORDOSIS.

Pseudopregnancy the reproductive state which occurs in many female mammals, such as rats and mice, after mating. Their cycles cease for a few days and only resume again if they are not pregnant.

Pupation The inactive state adopted during metamorphosis by those insects in which larva and adult are quite different. For example, maggots pupate and later emerge as flies.

Redirected activity behavior shown to an inappropriate stimulus when the appropriate one is not available. Denied food, a hungry pigeon may peck at pebbles.

Redundancy the inclusion in a message of more detail than is absolutely necessary for its recognition under ideal conditions. Redundancy in animal signals, such as their repetition several times, may make sure that the message conveyed by them is received and understood.

Reflex an automatic and involuntary response which is the simplest form of reaction to an external stimulus. Reflexes may involve as few as two or three nerve cells. Examples are a dog scratching when its side is irritated and the constriction of the pupil of the eye when light shines into it.

Regurgitation the method of feeding of another individual with food brought up from an animal's own stomach. Common in social insects, which share food in this way; and some birds, such as gulls, feed their young by regurgitation.

Reinforcement a term used by those who study learning for events which raise or lower the probability of a response. Thus, a rat may press a bar to obtain food (**positive reinforcement**) or to avoid electric shock (**negative reinforcement**). In both cases the animal is being rewarded.

Releaser any of the physical and behavioral attributes of an animal which have been molded by natural selection to elicit a response from other individuals. Most sounds and scents that animals make, as well as visual displays involving movements and bright colors, are examples of releasers (see also SIGN STIMULUS).

Resource holding potential the ability of an individual or group to monopolize resources, such as food or mates, and so deny others access to them.

Ritualization a process which occurs during the evolution of animal signals so that they become more stereotyped and striking, and hence clearly distinguished from the actions, such as DISPLACEMENT ACTIVITIES and INTENTION MOVEMENTS, from which they originated, as well as clear and unambiguous in the message they convey.

Rumination the system of digestion used by certain animals with cloven hooves, such as deer and cattle (**ruminants**). Newly eaten food is stored in a special stomach compartment, called the **rumen**, from which it is passed back to the mouth for chewing after eating is completed.

Rutting the behavior of male deer during the mating season when they herd females to form HAREMS and roar at and fight with intruding males. This period is often referred to as the **rutting season**.

Satellite male a male that associates with another male that is attractive to females; the satellite male attempts to intercept the females and mate with them as they approach. Satellite males attend territory owners on the LEKS of ruffs, for example, and also occur close to calling males among some frogs.

Scatter hoarding see HOARDING.

Scavenger an animal that eats organic debris and other decaying matter.

Scent marking the labeling of objects or individuals with scent, either from special glands or in the urine or feces. Scent marks function as signals and may transmit several types of information although their exact role in social communication is poorly understood.

Schooling the formation of fish into tight shoals, found especially in species subject to predation. Predators are confused and have difficulty in focusing on a single prey within a school.

Search image the phenomenon in which, although animals may have initial difficulty spotting cryptic prey, having found one item, they often quickly find others. They appear to have learnt to see the prey and are said to have formed a search image for it.

Sensitive period a period of time during development when animals are most likely to respond to particular stimuli. Many behavioral and physiological processes tend to occur at a specific stage because of this. The classic example is the IMPRINTING of young birds on their mothers which occurs most readily during the first few days of life.

Sensory adaptation the loss of sensitivity which occurs when sense organs are exposed for a long time to a strong stimulus. Sensory adaptation is one reason why animals may cease to respond to stimuli (see also HABITUATION).

Sign stimulus an external stimulus to which an animal responds in a specific way, such as the key features of a prey animal that elicit stalking or the swollen belly of a female fish that leads a male to court. Those sign stimuli evolved to function as signals are known as RELEASERS.

Social facilitation the effect whereby the performance of a behavior pattern by one animal leads others to start behaving in the same way or to show more of it. Unlike IMITATION it does not involve the acquisition of new behavior but is simply a way in which SYNCHRONY between animals in their activities is achieved.

Sociobiology that branch of behavior study concerned with the social behavior of animals, its ecology and its evolution.

Soliciting the posture adopted by a female which functions as an invitation to a male to mate. For example, LORDOSIS in mammals and a horizontal posture with tail fluttered up and down in many birds.

Song any lengthy sound signal produced by animals. Song may be simple, as in many insects, or elaborate, as in birds and whales. Most songs probably function in rival repulsion and mate attraction.

Sonograph machine for analyzing sounds which produces a plot of frequency against time known as a **sonogram.**

Sound spectrograph see SONOGRAPH.

Spatial learning learning the positions of objects in space. Spatial learning is very important for most animals so that they can become familiar with their HOME RANGE and the location within it of essentials such as food and shelter.

Spawning the production of eggs and sperm in those animals, such as most fish

species, where fertilization is external and eggs and sperm are just released into the water.

Specific hunger the occurrence of an appetite for particular nutrients. Animals needing salt will travel long distances to salt licks and, similarly, laying birds develop a taste for the calcium they need to make egg shells.

Stereotypy stereotyped and monotonously repeated movement patterns, especially those shown by caged animals such as the relentless pacing up and down of bears and lions in zoos and the tail-biting of farm pigs.

Stotting the adoption of a stiff-legged, bouncing gait by gazelles that have spotted a predator. This may function as a warning to others or possibly to the predator that, having been spotted, it has no chance of success.

Strategy mode of behavior adopted by an individual when it could, either actually or theoretically, behave in a different way. For example, males of some species may adopt either of two strategies, TERRITORIAL or SATELLITE. In theoretical discussions, **an evolutionarily stable strategy** is that which, if adopted by the whole population, cannot be bettered by immigrants adopting a different strategy.

Stress physiological state induced in animals by conditions they are unable to tolerate and cope with, such as pain or overcrowding. In mild form, the changes involved help the animal to cope but, if more extreme, they have adverse or even lethal effects.

Stridulation the method of sound production used by some insects in which they scrape together two parts of the body, such as a leg and a wing. The structures involved may be specially adapted to form a comb and a scraper so that a stereotyped rasping sound is produced.

Subordinacy describing the position of animals that lose in fights or defer to other individuals (see DOMINANCE).

Subsong quiet, less pure and less well-organized SONG that is produced by birds mainly early in the breeding season before they have developed their full song.

Supernormal stimulus an artificial stimulus which is more effective in

eliciting a behavior pattern than the normal stimulus found in nature. For example, many birds will attempt to sit on a huge model egg in preference to eggs of the usual size.

Surplus killing the habit of some predators of killing prey far in excess of their requirements. They do not then attempt to eat or to hoard the excess. This seems to occur mainly in occasional individuals confronted with large numbers of easily accessible prey, for example, a fox coming upon a gull roost on a night too dark for them to fly off.

Symbiosis a general term, used in North America for associations between species, including PARASITISM and COMMENSALISM, but limited in Europe to associations where both species benefit (MUTUALISM).

Synchrony the occurrence of the same behavior in different individuals at the same time. Synchrony may occur because both are responding to external cues, such as daylength or the presence of a predator, but sometimes it also arises through SOCIAL FACILITATION.

Taxis a movement in which an animal orients its body with respect to a stimulus, perhaps moving toward or away from it or taking a particular angle to it as in compass ORIENTATION (cf KINESIS).

Template a suggested sensory or neural recognition system that guides learning of SONG in young birds and ensures that they learn only those sounds appropriate to their species.

Territory an area occupied by one or more individuals of a species and defended against the intrusions of others (see also HOME RANGE).

Threat behavior shown in AGONISTIC situations which signals that the animal may attack. Dominants may threaten rivals who then withdraw, or an individual in a conflict between attack and escape may threaten instead of performing either of these actions.

Thwarting the situation where an animal is stopped from performing a behavior pattern by some external barrier or the absence of the expected stimulus. In such circumstances, as in CONFLICT, animals may perform DISPLACEMENT ACTIVITIES.

Tradition a behavior complex shared by individuals in a population and passed from one generation to the next through individuals learning from each other.

Trail following a response found in certain species, especially ants, in which workers lay PHEROMONE trails from the nest to sources of food, and other workers are able to reach the food simply by following the trail.

Trial and error a form of learning, also called instrumental CONDITIONING, in which an animal comes to associate performance of a behavior pattern with its consequences so that, for example, a response which by chance yields reward will be repeated more often.

Typical intensity the tendency of animal signals to be performed at the same intensity regardless of the MOTIVATION of the signaler. It is thought to have evolved through RITUALIZATION as a means to reduce the ambiguity of signals or to hide the motivation of the signaler where it is disadvantageous to indicate this to other animals.

Ultrasound sounds beyond the limit of human hearing (about 20,000 cycles per second). Such sounds are used in communication by many animals but, because they bounce back from objects, they are particularly effective in ECHOLOCATION.

Vacuum activity the occurrence of a behavior pattern in the absence of the external stimulus normally required for its elicitation, for example, prey capture when no prey organism is present, or courtship in the absence of a mate. Few instances have been described because the complete absence of a stimulus is hard to demonstrate.

Vigilance the state of an animal determining the likelihood that it will detect an unpredictable event in the environment such as the appearance of a predator. Feeding animals look up often to maintain vigilance, but less so in groups where the presence of more pairs of eyes makes it less necessary.

Viviparity embryonic development within the mother and receiving nourishment from her, as in all mammals and some other animals of various groups.

INDEX

A **bold number** indicates a major section of the main text, following a heading. A single number in (parentheses) indicates that the animal name or subjects are to be found in a boxed feature and a double number in (parentheses) indicates that the animal name or subject are to be found in a spread special feature. *Italic* numbers refer to illustrations.

O

P

Q

R

S

T